DESIGN FOR
3D PRINTING

SAMUEL N. BERNIER, BERTIER LUYT & TATIANA REINHARD

MAKER MEDIA™

SAN FRANCISCO, CA

Design for 3D Printing

by Samuel N. Bernier, Bertier Luyt, and Tatiana Reinhard

Copyright - © Hachette-Livre (Marabout) 2014

Printed in Canada.

Published by Maker Media, Inc., 1160 Battery Street East, Suite 125, San Francisco, CA 94111.

Maker Media books may be purchased for educational, business, or sales promotional use. Online editions are also available for most titles (http://safaribooksonline.com). For more information, contact O'Reilly Media's corporate/institutional sales department: 800-998-9938 or corporate@oreilly.com.

Editor: Brian Jepson

Interior Designer: Victor Grenez and Transparence

Cover Designer: Thomas Raygasse and Samuel N. Bernier

September 2015:   First English Edition

Revision History for the First Edition

2015-09-23:   First Release

See http://shop.oreilly.com/product/0636920032052.do for release details.

978-1-4571-8736-0

[TI]

DESIGN FOR
3D PRINTING

SAMUEL N. BERNIER, BERTIER LUYT & TATIANA REINHARD

SCANNING, CREATING, EDITING, REMIXING, AND MAKING IN THREE DIMENSIONS

MAKER MEDIA
SAN FRANCISCO, CA

BEFORE YOU START

To follow the exercises in this book, you have to download several free software packages, which will allow you to learn the basics of modeling for 3D printing.

On the **www.123dapp.com/create** website, you can easily download 123D Design and Meshmixer.

If you have an iPad, iPhone, Windows Phone, or Android phone or tablet, you can download 123D Catch from your device's application store. You can get 123D Sculpt for iPad on the App Store.

You will also have to download the Netfabb Basic software from **www.netfabb.com** website. This one will allow you to quickly cut and repair 3D files.

The free or paid version of Skanect will also be necessary for the 3D scan exercise, as well as a Microsoft Kinect (Xbox or Windows version, but not Kinect One), an Asus Xtion Pro, or Structure Sensor: **www.skanect.com**.

SUMMARY

FOREWORD BY CARL BASS

Maker and CEO, Autodesk

To paraphrase that old Hollywood witticism, 3D printing is an overnight success 30 years in the making…

Starting in the early 1980s, 3D printing has transformed from an experimental technique developed by a few early innovators—like Hideo Kodama and Chuck Hull—to a global movement embraced by millions of people all over the world.

For me, as someone who has always loved making things—boats, benches, baseball bats, and even rocket ships with my kids—it's been exciting to watch and even take part in this transformation; and today our company, Autodesk, is helping bring 3D printing to as many people as possible.

That's why I'm so happy to see a book about 3D printing written by true practitioners like Samuel, Bertier and Tatiana. In their FabShop in Paris, they do 3D printing every day, and they have a passion for teaching what they know to others. Their shop is always humming with activity and experimentation, and in the pages of this book they've captured that sense of excitement about doing new things with new technologies.

Inside you'll find a compelling depiction of the history of 3D printing, as well as lots of good "How To" sections and exercises to try. I think this combination of theory and practice will give the reader a deeper understanding of 3D printing to really learn what all the excitement is about.

I love the fact that 3D printing lets us do things we couldn't do before, and sometimes even things we couldn't conceive of before.

And I think that some of the readers of this book could be the ones to make the inconceivable, conceivable…

INTRODUCTION

Today, everyone's talking about it. To hear some manufacturers and journalists talk, you will have a 3D printer in your home in a few months or the next few years. Why wait for the future? For many people, there is already one available at your business, your school, your local library, or through online 3D printing services such as 3D Hubs.

All industries (at least those who produce physical things) use or will soon use 3D printing in their innovation processes: in the early development phase of prototyping as well as for the producing certain specialized or limited-run pieces, for which traditional methods of manufacturing are less adapted today.

At le FabShop, we use 3D printing every day: we set it up and teach our clients and the curious how to use it. We are "makers," new and connected DIYers who use 3D printing regularly, and we share some of our work on our website and on community websites.

This book project was born from our experience. If 3D printing ends up changing the world and, tomorrow, it is universally available at home or just down the street, then it needs to be made practical for the largest number of people.

To print an object in 3D, you need a 3D model and a 3D printer. To obtain a 3D model, there are three possible sources: 3D modeling software that lets you create an original model, a 3D scanner that allows you to virtualize a real object, or websites for downloading 3D files. We will explain later how to use one or the other of these solutions through practical exercises. We will also see what the major 3D printing technologies are, and which 3D printer to choose depending on the project and available materials.

If, at the moment 3D printing processes are still rather slow compared to industrial manufacturing processes, the future of these technologies is being played out today.

Dimensions, manufacturing time, available materials and the possibilities of combining materials are the areas that researchers, enthusiasts, and businesses around the world are working on.

It's time to get on this train because it's already on its way!

WHAT IS 3D PRINTING?

3D printing is a style of manufacturing technique known as an additive, as opposed to subtractive, process (even automated or robotized ones) that work by removing material. Subtractive techniques all have a certain number of limitations inherent in the methods of production. Indeed, to produce the final object, you need to start with a block of material, which causes waste and limits formal complexity.

Let's take, for example, the traditional way of manufacturing a wooden chair leg: a tree is cut down, the bark and branches are removed, and it is cut into boards. The boards, once dried, are planed, sawed, machined, and sanded. To make the same leg in plastic, you have to create a mold made from a block of metal, which the melted plastic will be injected into using a gigantic press.

Once cooled, the chair leg is ejected from the mold and deburred. Its injection "sprue," a sort of umbilical cord for the object, also needs to be removed. In order to be manufactured with this process, the leg of the chair needs to have a shape that meets the criteria of clearance angle, uniform thickness of the walls, and the other constraints of plastic injection molding.

All of these operations impose limits for designers when they are designing pieces. These manufacturing constraints are the reason why so many objects look and feel the same. For a designer, a new production technique, often means new aesthetic possibilities.

3D printing is a process that allows objects to be directly manufactured by depositing layers of material on top of one another. The layers of material deposited are usually less than a millimeter thick, or even less than a tenth of a millimeter with some technologies. There is a different technology that corresponds to each material. These techniques allow greater creative freedom and require fewer tools to create a prototype. The manufacturing time and cost for limited editions, or even single pieces, can also be reduced. What's more, the additive process makes it possible to manufacture objects comprising several pieces that are articulated in relation to each other in a single operation: in other words, you can pop a toy car out of the 3D printer, and it will be ready to roll with no assembly required!

Those are the principles and the advantages. Now let's look at the practice!

Next page:
Exhibition FABLAB / FABSHOP
at pavillon de l'Arsenal (march 2014)
Source: le FabShop
© Samuel N. Bernier

LEARN THE BASICS

ADDITIVE FABRICATION TECHNIQUES

These fabrication technologies all use the same technique: laying down consecutive layers of material to produce three-dimensional objects. Each of these digital prototyping techniques requires the intervention of a computer to interpret the digital models of the item to be printed. The digital model is *sliced* in hundreds of layers, each transformed into a path for the machine to follow while 3D printing. Certain technological solutions for the 3D printing of metal (DMSL, EBDM, DMD, SLM), of food (paste extrusion), and of body organs have not been included in this chapter. Several further techniques are also in the process of being researched, or are for industrial use only. Our selection in this list addresses the bulk of the rapid prototyping choices available to consumers right now either in the form of a service or in personal 3D printers. Pictures accompanying the text allow you to visualize the tooling and the operation of these machines. QR codes will direct you to videos that show each additive manufacturing process in action.

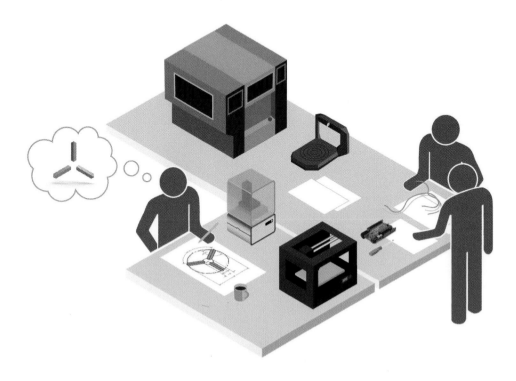

→ FDM AND FFF

Filament Deposition Modeling (Trademarked by Stratasys)
Fused Filament Fabrication (License-free terminology used by the open source RepRap project)

Inventor: S. Scott Crump (Stratasys).

Description: Melted filament deposition is one of the most widespread 3D printing techniques. The FDM technique was invented in 1988 when a man decided to manufacture a toy for his daughter using a hot glue gun. This man was no other than Scott Crump. This idea of automating the process of fabrication by depositing successive layers was the seed of the company Stratasys, which he created the following year with the help of his wife, Lisa Crump. Since 2009, the number of 3D printer models using this technique has exploded on the market, following the entry into the public domain of certain patents, among them Crump's. Not able to use the term FDM, trademarked by Stratasys, the creators of these new machines invented the term FFF, replacing the word "deposition" by "fusion." With the tooling and the consumables used by this technique being relatively affordable, simplified and democratized machines have been able to appear on the market. The RepRap project, for example, has allowed thousands of users to create their own open source FFF 3D printers. From this initiative was indirectly born the now-famous American brand MakerBot, which was then bought, in 2013, by Stratasys. The FDM process uses as its primary material thermoplastic filaments (a plastic that turns liquid under a certain temperature), which come wound on spools. These filaments, which generally vary in diameter between 1.75mm and 3mm, are mechanically pulled through an extruder heated to between 180 and 300 degrees celsius. The extrusion head acts like a hot glue gun and deposits molten plastic. This cools down almost right away in the ambient air. This way, consecutive layers of plastic are deposited in layer thicknesses of an average 0.2 mm, about as thin as human hair.

Some machines use a second extrusion head also, which allows support structures to be printed out of soluble material; such structures are there to support physical overhangs in the design of the main print. After printing is complete, the workpieces must then be plunged into a dissolving bath to remove those extra supports. Single head extruders can also build support structures, but in this case, the material used is the same as the main object and has to be removed manualy or with the help of sharp tools.

Materials: ABS (acrylonitrile butadiene styrene), PLA (polylactic acid), PC (polycarbonate), Nylon, PET (polyethylene terephthalate), PVA (polyvinyl alcohol) and, some experimental ones, like SWF (Seaweed Filament).

Manufacturers: Stratasys, MakerBot, Ultimaker, UP!, 3D Systems (Cube), LeapFrog, BQ, ShareBot, Be3D, PrintrBot, Pirate 3D, Solidoodle, Zotrax, Fabbster, Tinkerine…

Advantages: Choice of colors and materials, ability to create hollow and honeycombed volumes, wide range of prices, ability to create both solid and flexible workpieces.

Applications: Early product development, functional prototypes, "goodies", toys, robotic parts…

http://vimeo.com/77222355

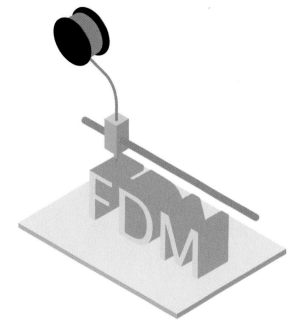

→ SLA
STEREOLITHOGRAPHY

Inventor: Chuck Hull (co-founder of 3D Systems).

Description: Stereolithography was the first 3D printing technology to be patented. This technology, attributed to Chuck Hull, founder of 3D Systems, used focussed UV light to harden a liquid photopolymer resin in order to create objects layer by layer.

Light-sensitive resin is put into a transparent tank while a laser, stationed above or below the tank, is directed using mirrors so that it follows a trajectory that tracks the shape sent from the computer. Once exposed to the strong light, the layers of resin are solidified, a phenomenon named *polymerization*, which causes them to stick together and produce a solid volume. Only one material may be printed at a time. The resolution of the layers that is obtained in SLA is on the order of 0.03 mm, which makes the technique highly-prized for its precision. Once finished, the workpieces must be plunged into a solvent bath, which serves to clean up the excess resin and to halt the chemical reaction. Created objects may complete their hardening in natural light, or else with the help of an ultraviolet oven. Objects made by SLA remain sensitive to the effects of UV radiation and may become damaged over time, by starting to show cracks or yellowing. However, the use of castable resins permits "lost wax" molding of some pieces. This characteristic is a major advantage for jewellers. The support structures (in the form of branches or 3D lattices) must be removed manually or with the help of tools. One evolution of this technology, DLP, uses a digital video projector in the place of a laser. This difference enables this process to cure the resin much faster and precisely by projecting light on entire layers in a single step rather than having to move the laser across the surface.

Materials: Colored and transparent photopolymer, flexible, thermofusible photopolymer.

SLA Manufacturers: FormLabs, 3D Systems, MiiCraft.

DLP Manufacturers: EnvisionTEC, Solidator, Prodways (with MOVINGLight technology), Nova 3D, Ember by Autodesk, Lightforge, The Deep Imager 5, Roland.

Advantages: Precision, transparency, fusibility, tiny workpieces.

Applications: Medical, model making, jewelry, figurine making.

http://youtu.be/NM55ct5Kwil

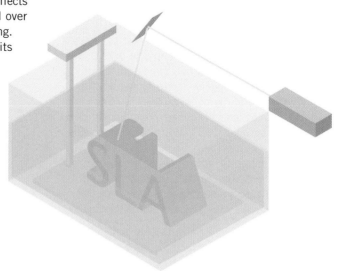

→ POLYJET
MULTIJET MODELING OR MATERIAL JETTING (MJM)

Inventor: Objet (Stratasys)

Description: The MJM technique resembles the functioning of an inkjet printer. With this process, a printing head moves around above a platform while jetting microdroplets of a photopolymer, a light-reactive resin similar to the one used in SLA. A UV light, situated near the print head, hardens the resin in place by curing (drying using UV radiation). By repeating this process, the MJM printer constructs a 3D object, one layer at a time. This technology also allows several different materials to be printed at one time, and for the materials to be combined. It is therefore possible to create workpieces in multiple colors, with multiple properties and opacities. With a layer resolution of down to 16 microns (0.016 millimeters), MJM 3D printers are considered high-end rapid prototyping tools. Their ability to reproduce the texture, appearance and functioning of items produced by injection molding makes them one of the preferred tools of industrial designers and engineers.

The ability to combine types of resins makes MJM technology the most versatile on the market, with more than 100 droplet layouts or arrangements available. The microscopic matrix composites thus created are called "Digital Materials." The support material for this process is gelatinous in texture and can be removed by hand or with the help of water jets.

Materials: Photopolymers which can be transparent, bio-compatible, colored, high-resolution, and/or simulate polypropylene, elastic or rubber, etc.

Manufacturers: Solidscape, 3D Systems, Stratasys (Objet).

Advantages: Large choice of materials and properties (rubberlike, transparent, etc.), colors, very high resolution (0.016mm), possibility of composite materials.

Applications: Medical, multi-material prototyping, injection molding effect, toy design, etc.

http://youtu.be/OpJonLX15PM

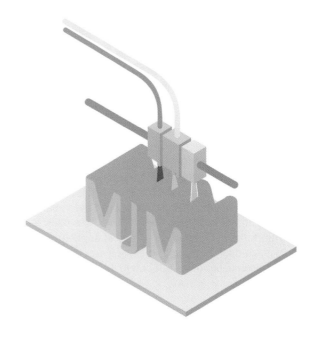

→ LOM
LAMINATED OBJECT MANUFACTURING

Inventor: Helisys Inc.

Description: This fabrication technique, invented in Japan, uses paper material as its primary consumable. The LOM machine unrolls a long piece of adhesive paper onto the printing platform, and then a heated roller applies pressure on the layer, which makes it adhere to the surface. Then, a blade or a laser cuts the contour of the paper to the dimensions of the platform, and traces the edges of the 3D object at that height. The unused space gets a checkered square pattern cut into it (these cuts will allow the object to be freed from the excess material around it when the fabrication process is complete). At each layer, the platform lowers itself by the thickness of one piece of paper material, and the sequence repeats itself until the workpiece is complete. At the end of the manufacturing process, you end up with a ream of glued-together paper that must be carefully pruned apart, in order to extract the object or objects that are hidden inside.

Given the volume that must be removed at the end in order to create a workpiece, LOM is more akin to subtractive, rather than to additive, manufacturing. The interiors of the pieces are full, because it is impossible to remove the support material imprisoned inside a closed cavity. This process, which had almost entirely disappeared from the market, made a stunning return with the arrival of the SDL technology (Selective Deposition Lamination), invented by the brothers MacCormack, creators of the Irish company Mcor. SDL replaces the proprietary adhesive rolls of paper with standard A4 sheets on which it is possible to print beforehand using an inkjet printer, in order to obtain objects with millions of colors. The glue is applied page by page using a robotic wheel. This is a process that uses paper to its full potential!

Materials: Paper and plastic film

LOM Manufacturers: Cubic Technologies, Solido.

SDL Manufacturers: Mcor Technologies.

Advantages: No chemical reactions, large undistorted workpieces, properties similar to wood (using paper), heat resistance, machinability.

Applications: Topographic maps or models, sculptures, structural, architecture, medical, archeological, museum, and scenery models, etc.

http://youtu.be/4ebj6hH0HnY

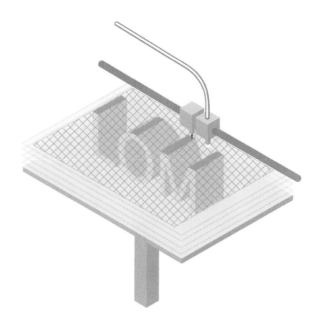

→ 3DP OR BINDER JETTING OR ZPRINTING

THREE DIMENSIONAL PRINTING, INKJET HEAD

Inventor: Z Corporation.

Description: This technique, developed at MIT (Massachusetts Institute of Technology), uses powders combined with colored binding agents to produce multicolored workpieces. Fine powders are added to a receptacle using a robotic arm that sweeps across the surface at a constant speed. With each pass, the binding agent is deposited in the correct areas through a nozzle, then the platform lowers by a few tenths of a millimeter in order to allow space for the next layer to be added. This sequence repeats itself multiple times until the print volume is reached. The printed pieces are then dug up out of the excess powder which is acting as their support, and cleaned using suction tools. Finally, the colored pieces are sprayed or plunged into a cyanoacrylate bath to give them greater durability and to achieve better color rendering.

Materials: Sandstone, plaster, sugar, acrylic powder, ceramic powder, calcium carbonate.

Manufacturers: Z Corp (1995-2012), 3D Systems (2012 until today), Foschif Mechatronics Technology.

Advantages: Color and freedom of shape design.

Inconveniences: Fragility, porosity, difficulty making sloping gradients with the cheaper versions of the machine.

Applications: Sculptures, figurines, reproductions, decorations, museum pieces, education, art.

http://vimeo.com/16882178

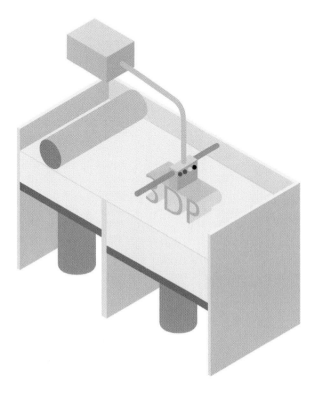

→ SLS
SELECTIVE LASER SINTERING

Inventor: Carl R. Deckard.

Description: SLS is a 3D printing technique invented in the 1980s, very popular for its ability to print complex and unique objects, or to produce a small batch of an innovative design. This technology uses a laser to "sinter" fine particles. Sintering consists of heating up a material to a temperature just beneath its melting point, in such a way that its particles agglomerate with the neighboring particles. Under the effects of heat, the powders targeted by the laser flow together, thus forming the cohesive piece. The grain size of the consumable is therefore very important because it will partly define the resolution that will be possible to achieve in the finished piece. The powder may have a thermoplastic or metallic base, depending on the strength of the laser that is used. While the SLM (selective laser *melting*) process allows powders to be used in their pure state, the SLS technique often resorts to using composite or mixed powders in order to facilitate the consolidation of the workpieces. Fabrication takes place in a tub of encapsulated powders in a controlled environment. To start a print, a robotic roller spreads a fine layer of material across the printing platform. The laser passes over the covered surface and sinters the powders according to a cross-sectional perspective of the object. The powder/laser sequence repeats itself while the tub progressively fills. The excess powder is used as a support for the printed object, and can be recycled for a future print. At the end of the fabrication cycle, the controlled-environment chamber is allowed to cool down, and the excess raw material is removed in order to reveal the pieces and clean them (often, an air jet is enough). The objects thus produced using a nylon powder base are porous and can easily be dyed in a coloring bath, or even polished in a tumbler. SLS printer vats can be optimized to print a large number of projects at a time, which makes this the preferred tool of on-demand 3D printing services. Note that laser sintered models are often hollowed to allow for a better optimization of the machine's consumable materials. For this reason, holes have to be created on some models to enable the removal of excess powder.

Materials: Nylon (sometimes combined with other "non-sinterable" materials), polystyrene, steel, titanium, "green sand," etc.

Manufacturers: EOS, 3D Systems, Voxeljet.

Advantages: Speed, complex workpieces without extra support, robustness and flexibility, possibility of large-format prints.

Applications: Architecture, commercial printing services, design schools, prototyping.

http://vimeo.com/14737152

http://youtu.be/YN8NZJYboHg

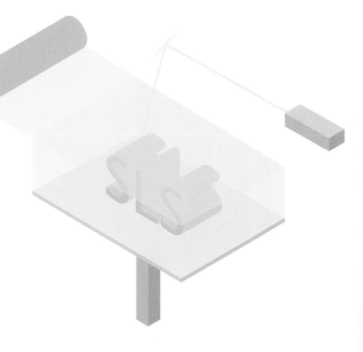

CHOOSING A PERSONAL 3D PRINTER

As mentioned in the previous section, there are several technologies and different price ranges involved in the field of 3D printing. FDM/FFF was the first technology to be made more accessible to the public, so we'll list the different brands using this technology and what differentiates them in this rapidly evolving market. Therefore, we describe here the origin of some innovative companies and how they contributed to the democratization of digital manufacturing. We also list a few 3D printing brands offering relatively affordable SLA and DLP solutions.

MAKERBOT

This American 3D printing brand, created in Brooklyn in 2009, is a result of the joint effort between three makers: Bre Pettis, Adam Mayer, and Zach Smith. This group's first machine was known as the now legendary Cupcake, a small ABS filament 3D printer with laser-cut wood structure and whose melted plastic layers would, with some difficulty, reach 0.3 mm thickness. It was very closely followed by the Thing-O-Matic, then by the Replicator, still made of wood but of much larger size. MakerBot's true shining moment came with the launch of the Replicator 2, in September 2012. The wooden casing was replaced by a solid steel structure, PLA filament was made available instead of ABS and the heated aluminium printing bed got replaced by a thick bed made of acrylic.

From then on, MakerBot would develop its own environment, without help from the open source community, launching proprietary software known as MakerBot Software, replacing the previous Replicator G program. The Replicator 2 long held number one position at the top of a now wide range of 3D printers costing less than $3000. In 2014, right after the firm Stratasys acquired Makerbot, competition became fiercer with machines from the entire world sold at competing rates. The models following the Replicator 2 haven't yet turned out to be as successful, whether it be the Replicator 2X (technologically reverting to the original Replicator with two extruders and a heated printing bed) or the Replicator 5th generation (providing an integrated monitoring device, a glass build plate, WiFi, extruders with an interchangeable mounting system and LCD display).

The summer of 2014 saw two new incoming products: a small size Replicator Mini Desktop 3D printer and an extra-large Z18 model. MakerBot has not yet had its final say and remains one of the main innovators in the 3D printing industry.

Name : Replicator Mini
Country : USA
Size (mm) : 295 x 310 x 381
Build volume (mm) : 100 x 100 x 125
Filament diameter (mm) : 1.75
Min. layer thickness (mm) : 0.20
Material : PLA

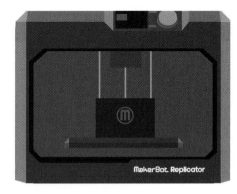

Name : Replicator (5th génération)
Country : USA
Size (mm) : 528 x 441 x 410
Build volume (mm) : 252 x 199 x 150
Filament diameter (mm) : 1.75
Min. layer thickness (mm) : 0.10
Material : PLA

gained momentum. It wasn't until 2008 that the RepRap project became an international phenomenon and that the release of the RepRap Darwin was announced. All around the world designers, engineers and hobbyists had been dreaming of an affordable additive manufacturing solution for almost two decades. They were far from imagining that it would take the form of a kit you assemble yourself. Plans for building RepRap 3D printers became available for free on the internet, and the variety of machine models grew rapidly.

Even today, despite the dozens of firms offering affordable 3D printers entering the market, RepRap continues its development. After Darwin, the Mendel and the Huxley printers were launched, but a substantial amount of derivatives also exist, coming from ingenious designers all around the world. Moreover, the term *RepStrap* was coined in order to define 3D printers based on the RepRap project but produced using other methods, such as laser cutting for making the frame. The first MakerBots, as well as the current Ultimakers and PrintrBots fall into this category.

Name : Replicator Z18
Country : USA
Size (mm) : 493 x 565 x 854
Build volume (mm) : 300 x 305 x 457
Filament diameter (mm) : 1.75
Min. layer thickness (mm) : 0.10
Material : PLA

REPRAP

In 2005, Doctor Adrien Bowyer, a professor at the University of Bath, in the United Kingdom, developed the following idea: to design a rapid prototyping machine capable of "reproducing itself", by generating the pieces needed for the creation of a "clone". Fused deposition modeling was a perfectly adapted technology. Several key FDM patents had fallen into the public domain and Scott Crump's invention was about to experience an unexpected revival beyond the reach of its original creator. The RepRap project (a contraction of Replicating Rapid Prototyper) started off as a blog in March of the same year.

The open source 3D printing revolution had begun and the Maker Movement in the United States thus

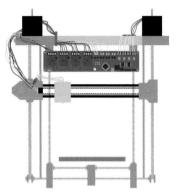

Name : RepRapPro Huxley
Country : United Kingdom
Size (mm) : 260 x 280 x 280
Build volume : 140 x 140 x 110
Filament diameter : 1.75
Min. layer thickness (mm) : 0.20
Material : PLA, ABS

ULTIMAKER

This small company from the Netherlands, created in 2011 (three years after MakerBot), is iconic among 3D printing firms. Ultimaker also spawns from the RepRap open source project. In contrast to Adrian Bowyer's project, the Ultimaker is not meant to "self replicate". Similar to its American competitor, Ultimaker developed its trade by selling a wooden 3D printer in the form of a DIY kit (the Ultimaker Original), but in contrast with MakerBot, the firm decided to pursue an open source development. This strategy may have affected this firm's search for investors and the subsequent growth of the company, but enabled the firm to maintain a high image. This decision's positive commercial return has been a constant sympathy emanating from FabLabs and Makerspaces all around the world. In 2013, the Ultimaker 2 was launched. It came factory assembled. Its elegant design and professional look made this printer successful enough to become the main competitor of MakerBot's Replicator 2. Like many others, this 3D printer has cube-shaped format, but uses a software named Cura (created by the Ultimaker community) to slice the 3D model, and uses 3 mm filament instead of the more common 1.75 mm. With its heated glass bed, it is possible to print ABS or PLA. The Cura software offers numerous possibilities in choosing printing parameters, which experienced users find greatly appreciable. The firm has also developed a web-based platform named YouMagine, making it possible to share, download and comment on others' 3D files.

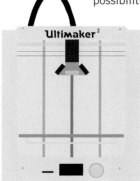

Name : Ultimaker 2
Country : Holland
Size (mm) : 350 x 350 x 400
Build volume (mm) : 230 x 225 x 205
Filament diameter (mm) : 3
Min. layer thickness (mm) : 0.05
Material : PLA, ABS

ZORTRAX

This Polish brand of 3D printers came as a surprise in 2014. When the whole world was expecting 3D Systems, Stratasys, or PP3D to develop the next machine that would combine reasonable pricing, higher resolution and elegance, this young group, then unknown to the world of 3D printing, offered a new machine that became a new market leader in Europe. Their first M200 model, with its proprietary plastic spools, metallic frame, heated printing bed and dedicated software is able to challenge the very best models. Its price, well below other printers, will help the brand develop among young professionals and hobbyists alike. The M200 creates impressively accurate ABS parts, while its software, Z-suite, generates one of the best support material an FFF 3D printer could get. Everything seems to have been thought out, in order for the firm to stand out from a market already saturated with competitors, in regard of brand image, of its well-polished communication campaigns, and even the high-end aspect of the products offered.

Name : M200
Country : Poland
Size (mm) : 345 x 360 x 430
Build volume (mm) : 200 x 200 x 185
Filament diameter (mm) : 1.75
Min. layer thickness (mm) : 0.09
Material : ABS

FORMLABS

This company was historically the first one to offer an affordable stereolithography (liquid resin) 3D printer. Its significant success on Kickstarter in October 2012, with nearly 3 million dollars raised, has turned this MIT students' newly co-founded company into one of the symbols the third industrial revolution.

Benefiting from the expiry of certain key patents linked to the oldest rapid-prototyping technique, the Form 1+ model offers the best layer resolution in its price range. Formlabs CEO and co-founder, Maxim Lobovsky, like majority of tech company leaders, found a few bumps on his path, such as lawsuits from 3D Systems, the inventors of the SLA technology. It didn't stop the company from releasing a remarkable little machine, perfectly adapted for jewelry designers and makers of small components. Along with an extremely simple use of proprietary software and a finishing kit which is essential for this technology, the Form 1 is a gem in the 3D *Desktop* printers category. Superbly designed, with an extremely silent print and incredible precision, only three limits prevent it from being the perfect prototyping tool: the high cost of resin, the low mechanical properties of printed pieces, and the small volume of production. The Form 1+, a faster and more precise version of the original

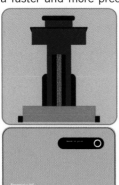

Formlabs machine, is now the leading product of the brand.

Name : Form 1+
Country : USA
Size (mm) : 300 x 280 x 450
Build volume (mm) : 125 x 125 x 165
Min. layer thickness (mm) : 0.025
Material : Photopolymer resin

EMBER (A.K.A SPARK)

In May 2014, when Autodesk announced SPARK, an open platform to help create better software, materials and services dedicated to 3D printing, the maker community got very excited. But what got the media's attention was the beautiful, cylindrical, and open source DLP (Digital Light Processing) 3D printer .

Autodesk's first piece of hardware was intended to showcase the repair, slicing and toolpath generation of SPARK, but it became a mystical piece of equipment that everybody wants but that very few people get to even try. The machine was even nameless for a few months, which brought some people to call it SPARK, referring to the platform. Today, some lucky makerspaces, mostly in the US, can brag that they own an Ember, the unicorn of 3D printers, faster and more precise than anything you have seen before.

Name : EMBER
Country : USA
Size (mm) : 325 x 340 x 434
Build volume (mm) : 64 x 40 x 134
Min. layer thickness (mm) : 0.025
Material : Photopolymer resin

PRINTRBOT

Printrbot is a project created by Brook Drumm, a California web designer. The Kickstarter campaign for the launch of his first 3D printer model encountered immediate success in 2011. By borrowing a lot of ideas from the RepRap community, he managed to create an expandable machine at a reasonable price.

Since his first triumph, seven other models were added, slowly switching from laser cut wood frames to metal ones. The Printrbot products aim at customers of all ages who are discovering the 3D printing field. These small tools that can be assembled in a matter of hours are a perfect fit for home use (weekend projects, repairs, creating small spare parts...). Figuring among one of the least expensive ranges of printers, this brand truly contributes to widely spread the use of 3D printing at home. Printrbot, not having a proprietary software, relies on popular slicers such as Repetier-Host or Cura (for the most recent machines).

This brand, with the outstanding simplicity of its products, is perfectly in line with the Do-It-Yourself and Maker movements. This firm's ultimate goal is to provide every home and school with a 3D printer.

Name : Printrbot Simple Metal
Country : USA
Size (mm) : 356 x 254 x 406
Build volume (mm) : 150 x 150 x 150
Filament diameter (mm) : 1.75
Min. layer thickness (mm) : 0.10
Material : PLA

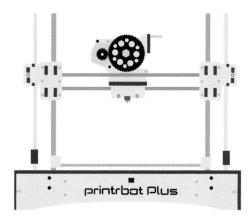

Name : Printrbot Plus
Country : USA
Size (mm) : 508 x 508 x 508
Build volume (mm) : 203 x 203 x 203
Filament diameter (mm) : 1.75
Min. layer thickness (mm) : 0.10
Material : PLA, ABS

CUBIFY

Cubify is a product line developed by 3D Systems in order to enter the market of private consumers. Very shortly after its launch, the Cube desktop 3D printer could be found on the shelves of consumer electronics stores around the world. This stroke of marketing genius was an important step towards spreading the use of 3D printing, but it also left those who thought they had purchased an all-purpose machine with bitter memories.

The Cube is considered by many as a "rich person's toy" since it isn't precise enough for professional use, nor is it affordable for most hobbyists. Furthermore, basing its model on ink printers, the Cube can only be used with non-refillable cartridges which are far more expensive than those sold by MakerBot and other companies. The original Cube printer had a design similar to PP3DP's Up! model with a plastic enclosure rather than metal and with more rounded edges.

Its large-sized version, named Cube Pro Trio, can use up to three colors at the same time, unfortunately, this impacts the generated pieces' quality. However, the Cube is one of the only 3D printers that can be used without danger by a child and to offer multiple advantages, such as a magnetic printing bed and Wi-Fi connection. The tools developed by Cubify were up until now disparaged by the 3D printing community, but this trend may be reversed when their 3D scanning tools such as Sense and iSense are taken into account. Also, a special edition of the Cube, named EKOCYCLE, uses plastic partially recycled from Coca-Cola (polyethylene) bottles, an eco-friendly as well as an astute marketing approach.

With a brand image matching current tastes, and solutions avoiding past mistakes, 3D Systems, the 3D printing giant has regained some ground in today's personal 3D printer market. The consumer market is approaching a maturity that draws a real benefit from additive manufacturing—Cubify will probably become an important partaker of this homemade revolution.

Name : Cube 3
Country : USA
Size (mm) : 335 x 338 x 280
Build volume (mm) : 152 x 152 x 152
Filament diameter (mm) : 1.75
Min. layer thickness (mm) : 0.10
Material : PLA, ABS (recycled PET)

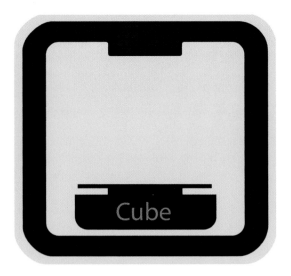

Name : Cube Pro
Country : USA
Size (mm) : 578 x 578 x 591
Build volume (mm) : 273 x 273 x 241
Filament diameter (mm) : 1.75
Min. layer thickness (mm) : 0.07
Material : PLA, ABS

PP3DP

With its small, metallic painted and colorful frame, its aesthetic design and impressive speed, the first personal and transportable personal 3D printer from the Chinese manufacturer PP3DP had everything you could ask for. Several generations of printers later, the UP! Plus 2 has a lot in common with its "forefather", but with even more rapidity, precision and connectivity. When it came out in 2010, the UP! 3D printer was greatly criticized for its exclusive "plug and play" mentality. Very far from the open philosophy of the RepRap project, the firm PP3DP produces machines with hardware and software that excludes all forms of improvement, configuration, or repair. It is thus extremely complicated to modify the printing bed or extruder temperature, to replace a component on the motherboard, or to adjust the speed of the machine.

Although it may annoy hackers, makers and other tinkerers, the UP! 3D printer works well and the results are sometimes impressive. It is very popular among schools whose first preoccupation isn't to explain the mechanisms involved, but to produce pieces designed in a CAD program. An even smaller version is also available called the UP! Mini, whose mechanism is entirely covered by a shell in order to maintain the ambient heat while printing. The preferred printing material used with UP! is ABS, a capricious consumable when it comes to large prints since it tends to bend from retraction caused by the cooling process. However, ABS is unexpectedly well-handled by this machine and that's partly why UP! is so well appreciated for the fabrication of small objects with good mechanical properties. It is possible to use 1.75 mm plastic filaments coming from brands other than PP3DP, but considering their melting temperature they might not offer the same level of finishing as the official consumable.

Name : UP! Mini
Country : China
Size (mm) : 240 x 355 x 340
Build volume (mm) : 120 x 120 x 120
Filament diameter (mm) : 1.75
Min. layer thickness (mm) : 0.15
Material : ABS, PLA

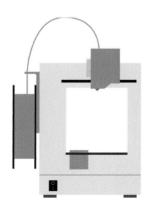

Name : UP! Plus 2
Country : China
Size (mm) : 300 x 260 x 350
Build volume (mm) : 140 x 140 x 135
Filament diameter (mm) : 1.75
Min. layer thickness (mm) : 0.15
Material : ABS, PLA

TINKERINE

This Canadian company has nothing to be sorry for. After the success of their wooden DITTO 3D printer, the Vancouver-based team came up with a shiny and minimalist new design to attract designers. We could say that Tinkerine is the Ultimaker of North America. Not only for its design, but also for the strong community they managed to build over the years. The Ditto has a characteristic C-frame that makes the build platform very accessible and easy to keep clean, unlike the closed box that most FFF 3D printers have. The Ditto Pro also has a hidden spool compartment, making this sophisticated machine fit on an office desk without looking awkward. The print quality is impressive and so is the Tinkerine Suite software: an intuitive interface with Cura's power under the hood. Just like Ultimaker 2, the Ditto Pro has a glass build plate that may need to be covered with tape or glue to prevent parts from warping during the print.

SOLIDOODLE

Solidoodle was created by an American engineer, formerly employed by Makerbot in Brooklyn as Operations Manager. With his strong experience, he was able to provide the market with a pre-assembled 3D printer at a competitive price. This firm's immediate success which came as soon as the announcement was made that an FFF machine costing under $500, with no assembly required, would be placed on the market. At the time (2011), the only 3D printers available were sold as do-it-yourself kits that the user would find very painful to build properly.

Nevertheless, the original model was sold "bare bones," with no enclosure and no unnecessary accessories. To "dress it up", a small extra fee was needed. As newer printer generations came by, the price of the Solidoodle has increased little by little (and also in size and quality of print) to reach $1,000. At such a price, other brands may be preferred but the Solidoodle remains an open and relatively simple-to-use 3D printer. Some components (the same applies to most 3D printers) have to be readjusted once in a while.

While the Solidoodle is capable of printing PLA, it is originally optimized for ABS consumables, without having to buy proprietary refills. If you are not the least bit resourceful, this 3D printer is

not for you. But it is a favorable option for the designer or engineer on a tight budget.

Name : Ditto Pro
Country : Canada
Size (mm) : 370 x 390 x 436
Build volume (mm) : 225 x 165 x 205
Filament diameter (mm) : 1.75
Min. layer thickness (mm) : 0.05
Material : PLA

Name : Solidoodle (4th génération)
Country : USA
Size (mm) : 343 x 356 x 381
Build volume (mm) : 203 x 203 x 203
Filament diameter (mm) : 1.75
Min. layer thickness (mm) : 0.10
Material : PLA, ABS

BEEVERYCREATIVE

This brand is known for its highly portable and good looking filament 3D printer, BeeTheFirst. Based in Portugal, the company went for a "plug and play" type of machine intended for designers, schools and consumers. Awarded across the world for its clever design, the Bee may not be the best 3D printer you will find in this list, but definitely deserves to be mentioned. While in the same category as the MakerBot Replicator Mini, using small PLA spools with an average layer resolution and slow speed, the Bee also has some very interesting features such has a handle that collapses to fit perfectly with the printer's frame, a magnetic build plate and great out-of-the-box experience. Instead of filling the machine with useless gadgets, BeeveryCreative decided to solve their problems in the most simple ways magnets, mechanical adjustments and a strong metal body. There is no LCD screen either, everything is controlled from BeeSoft, which you install directly on your computer. One downside of the smooth design is that it is difficult to clear the extruder when the nozzle gets clogged, because the filament is laborious to access once it passes into the enclosure.

DREMEL

Dremel is a company loved by makers and tinkerers for their affordable and widely-available manual tools. It only made sense that this brand, which is over eighty years old, would try to open to a more digital world. This American-born company, now owned by BOSCH, knows how to please consumers and how to keep a good brand identity. Their first 3D printer, the Idea Builder, is based on the FlashForge Dreamer, a Chinese 3D printer based itself on the MakerBot Replicator 2. Dremel brought what they do the best to their original machine: user experience, documentation, good prices and know-how. Being Dremel gives them the opportunity, more than any other, to be sold in hardware stores as the ultimate DIY tool. The Idea Builder is not targeted for professional designers but is a good start for any person looking for a new fancy toy to equip their garage. Just like most brands aimed at consumers, Dremel created a simplified software meant to facilitate the preparation of 3D files. They might have removed too many functions for their own good, but once again, the Idea Builder is meant to provide a simple and easy-to-use 3D printer to the masses at an attractive price point.

Name : BEETHEFIRST
Country : Portugal
Size (mm) : 400 x 140 x 400
Build volume (mm) : 190 x 135 x 125
Filament diameter (mm) : 1.75
Min. layer thickness (mm) : 0.10
Material : PLA

Name : DREMEL 3D Idea Builder
Country : USA
Size (mm) : 400 x 485 x 335
Build volume (mm) : 230 x 150 x 140
Filament diameter (mm) : 1.75
Min. layer thickness (mm) : 0.10
Material : PLA

IDEA SERIES (STRATASYS)

Stratasys is an unquestioned leader in the field of 3D printing. Owning multiple patents around fused deposition modeling (FDM) technology, the international corporation is also the owner of numerous companies around the world, including MakerBot.

Although it is a firm well recognized for its high-end equipment aimed at large companies and manufacturers, Stratasys is also the processor of the "Idea" series, which includes the famous Mojo printer. This small professional rapid prototyping machine holds many features usually associated with industrial additive manufacturing equipment, notably in terms of dissolvable support material, generated by a second extruding nozzle. The minimum layer thickness allowed by the Mojo printer is inferior to all other machines mentioned earlier, yet this does not excessively affect the surface of the pieces produced, due to a controlled environment and the use of proprietary consumables. The cost of use of this machine is much higher than other printers in the present list, but this does not prevent it from representing a satisfactory compromise between higher quality solutions (which can cost several tens of thousands of dollars) and entry-level printers.

The use of dissolvable support materials is one of the keys of fused deposition 3D printing. Without this removable material, certain geometries become more complex to produce, especially when it concerns moving parts. This advantage also has a price: time. Indeed, in order to dissolve the support, the piece needs to be immersed in an appropriate solution for several hours, thus extending the time needed to obtain a prototype. There is no machine currently belonging to the "prosumer" or "personal" range of printers capable of building dissolvable supports as well as the Mojo, despite the great number of 3D printers with double extruders on the market. Yet, this is not due to the lack of experimentation with the noteworthy dissolvable PVA and HIPS materials. Stratasys seems to have more than one trick to teach to the hackers belonging to the open source movement.

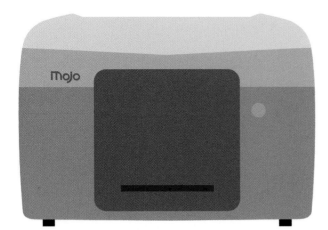

Name : Mojo
Country : USA
Size (mm) : 630 x 450 x 530
Build volume (mm) : 127 x 127 x 127
Filament diameter (mm) : 1,75
Min. layer thickness (mm) :0.17
Material : ABS

FILE FORMATS FOR 3D PRINTING

There are many file extensions used in the world of 3D modeling. Most 3D software uses its own proprietary format, such as SKP files for SketchUp, SLDPRT for SolidWorks, etc. Some software, like Blender, is more flexible with the file types accepted when importing, and also for exporting. In this ocean of proprietary formats, only a very few can be directly used with 3D printing software. Fortunately, some 3D file formats have distinguished themselves over time, becoming a standard for digital manufacturing. These are the "generics" of 3D, files that can be read and exported by most types of software, either by adding a plug-in, or directly supported in the software's import or export options. The STL file format reigns supreme in the world of digital manufacturing, but other file extensions are becoming increasingly popular, especially with the arrival of multicolor 3D printing. The three-dimensional mesh of an object defines, in part, the resolution that the printed object will have. The triangles that make up this framework are the equivalent of pixels in digital images.

In order to be usable for additive manufacturing, a three-dimensional geometry must be *closed*. This means that each side of the object's mesh must belong to two triangles at a time. Even though some software deliberately ignores this rule when exporting the 3D file, others will automatically prevent you from creating a file if it is not considered "watertight". Most of the time, the mesh generated by CAD software will not be used as-is by the additive manufacturing machine. Instead, the 3D information in it will be translated into a series of 2D cuts (one for each layer of the object), either in tool path code (GCODE) or into another proprietary file incorporating the other information necessary for the machine to operate (temperature, speed, power, etc.). The following file formats act as transitional languages between the design software and the software controlling the production tool.

→ STL

This file format was created by the 3D Systems company as a native extension of their Stereolithography software, used for the rapid prototyping technique of the same name. Also known as Standard Tessellation Language, this format is common to most additive and subtractive manufacturing tools. It is by far the most used of all file formats for 3D printing. STL meshing represents the geometry of a three-dimensional object, without any reference to its color, material, or texture. When importing an STL file in a 3D software, it is therefore "amnesic". It retains no memory of the functions that helped to shape it and can also lose trace of its original unit of measurement. For this reason, it is important to check carefully that the scale of the object is consistent with that of the original model. It's common to see an STL file imported at a scale of 1:100 in modeling software or switching from a metric sizing to its imperial equivalent (inches). Some software allows you to choose which unit of measurement in which to import an STL file.

When exporting an STL file, it is possible to choose the density of the meshing. This resolution will partially define the surface state that the part will have once manufactured, but will also have a direct impact on the "size" of the file. There are two types of STL files: ASCII and Binary. The second one is the most used because it's more compact for an equivalent resolution.

→ OBJ

This file format was developed by Wavefront Technologies for an animation software package. For this reason, some old-school computer specialists tend to call this type of extension a *Wavefront file*. As an open format, OBJ has been adopted by many 3D graphics design solutions. There are only a few 3D software programs that do not accept the exporting or importing of files in OBJ format. Just like STL, this export simply represents a geometry, without any information on the unit of measurement, or on the stages in the modeling process. However, OBJ has two great advantages, making it a popular format among graphic designers. Firstly, it accepts *NURBS-type* surfaces, making it possible to define complex curved surfaces without the need for an extremely dense mesh of triangles. Secondly, OBJ is supplemented by one or more *MTL* (Material Template Library) files, containing the information on color and texture linked to the 3D object. MTL is another standard created by Wavefront which references materials in a library of textures. This companion file makes it possible to view multicolored files and manufacture them by factoring in their chromatic complexity. Of course, transparencies and reflections will not be reproduced in 3D printing.

→ VRML

This graphics file format, developed for the World Wide Web, has long been superseded in the world of animation by formats like X3D, which are a lot more precise for photo-realistic renderings. However, since 3D printing only needs basic graphics information, WRL files are still being used for scanning as well as for manufacturing. Virtual Reality Modeling Language uses the file extension WRL, which refers to the word "world". This international standard has never been aimed at additive manufacturing, but its use of three-dimensional polygons combined with UV textures is suitable for the specific use of color 3D printers, like the Mcor Iris paper 3D printer. The terms VRML97 or VRML 2 indicate the most recent version (1997) of the file, but they don't have much to offer when it comes to manufacturing. Even though the WRL extension never had its anticipated success for the internet, it continues to be a standard in 3D file exchanging.

→ COLLADA

The COLLADA format was created by Sony Computer Entertainment to offer a file format that's easily shareable between different types of 3D software and accessible to a wide audience. COLLADA is now managed by a non-profit consortium called the Khronos Group. Using the DAE (Digital Asset Exchange) extension, this file format is still considered as an intermediary, used to transport graphics information from one software program or 3D app to another. Thanks to its popularity in the world of software interchange, COLLADA has also found a niche for itself in some apps designed for 3D printing, making it possible to import a DAE in the same way as an STL.

⇢ PLY

This file format is often seen in the world of digital manufacturing because it's mainly used to contain information captured with a 3D scanning tool. Like the other formats we mentioned, it makes it possible to describe a three-dimensional volume with polygons, and it comes in both ASCII and Binary versions, as does STL. PLY was created by students at the Stanford Graphics Lab to compensate for certain weaknesses in the OBJ format. Also known as Polygon File Format, PLY can contain information about color, transparency and texture, which means it's a format used mainly for graphics. It can easily be transformed into a simple STL file using software like MeshLab.

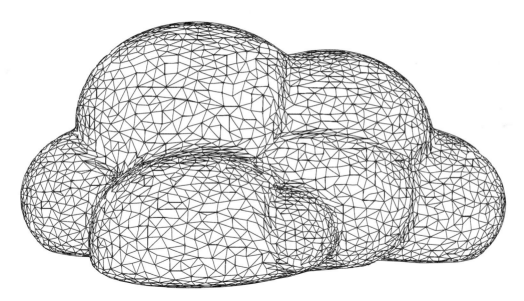

In 3D printing, volumes are defined by polygons.
Here is the cloud shown on page 61.

THE 3D PRINTING FILE CHECKLIST

→ COMPATIBLE FILE

STL for the most 3D desktop printers, OBJ or WRL for files with texture. Some 3D printing software will also read X3D, DAE, PLY, or COLLADA file formats.

→ POLYGON LIMIT

Avoid files with extremely dense meshing. 1 million polygons should be your maximum threshold. Try to generate STL files that are no larger than 50 Mb. Beyond that, some printing services and software could have problems managing them.

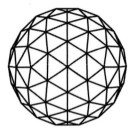

→ ORIGIN

When designing your model, it is better if the XYZ origin of your model's workspace touches a corner of the model, or is in the center of the volume. If not, you risk the object appearing outside of the virtual build area when it is imported into the slicing software (Repetier-Host, Cura, etc.). It is then situated far from the virtual space shown on the screen and you'll have to move it with the Move tool. You can ensure that the model will import correctly by snapping your 3D model to its environment's origin (X=0, Y=0, Z=0) before you export it as a 3D printable mesh.

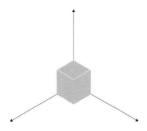

→ COHERENT 3D MODEL

Your model should be watertight, without volume collision or double surfaces.

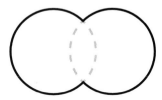

→ SUITABLE MATERIALS AND TECHNOLOGY

Some of the details of your model may vary in PLA, in paper, or in ceramic. The minimal wall thickness, shrinkage, the choice of finish and layer resolution will change depending on the material and technology used. It is essential to research the individual characteristics of each 3D printer and to choose it according to the type of objects that you need to produce.

→ EXHAUST PORTS

If you're planning to use powder (SLS, 3DP) or liquid resin technology (SLA, DLP), you have to provide ports which will allow substances to be released from hollow bodies. If not, the powder or resin will be trapped between the solidified walls. This tip also saves time and materials.

→ SCALING

It's better to check the scale of the model to be printed before exporting it. When you load it into other software, it may ask for the unit of measurement used. In some cases, STL files modeled in millimeters can appear in meters or even in inches when you load it into your 3d printer's software. You'll have to restore the correct ratio using your software's Scale option to return them to their initial scale before manufacturing.

→ PRINTER SIZE

Remember the maximum sizes of your manufacturing tool before exporting your file. If your object is too big, you'll have to cut it into several sections or choose another machine. For example, If your object exceeds a cubic meter, consider producing it using a subtractive method, like digital milling (CNC).

→ DETAILS

Make sure that the details, embossing, text, and textures stand out enough to be visible or readable. 2D images and letters attached to a file won't appear on a monochrome 3D print, unless they are transformed into a "bump map", often used to create complex 3D textures in animation and sculpture software.

→ TOLERANCES

If you have assemblies, sockets, or movable parts, make sure you have enough play to allow for the movement of the different bodies. Several trial runs may be necessary in the case of mechanical parts, but, in most processes, a minimum space of 0.25 mm is necessary to prevent the parts from fusing while being built.

→ FILE NAME

Get into the habit of naming your files according to a standard procedure. Firstly, never use special characters or accents and replace spaces by underscores. Choose a name that describes the content of the file clearly, its version, its scale, and any other information that will help you avoid having to open the file needlessly.

sphere3_10X10.stl

→ ORIENTATION

It is vital to choose carefully the orientation in which your part will be printed. This choice will have a great impact on the solidity, surface state, and printing time. If you are using a printing service, they will determine the orientation in which your object will be printed.

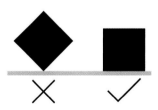

FREE 3D MODELING SOFTWARE

There is a huge number free 3D modeling software programs. It's important to choose yours wisely according to the project you're working on. Here is the software we recommend for your first experience of 3D modeling.

TINKERCAD

TinkerCad is the perfect 3D modeling software for beginners. Its 3D modeling method is like working with LEGO bricks. Except, those bricks can be customized and even subtracted from. It has fun tutorials to learn how to use the interface and understand the construction logic. It is ideal for family or school use. This software is only available online, so make sure you have an internet connection when you need to use it.

SKETCHUP MAKE

MAKE is the free version of SketchUp. You can use most of the tools that are in the Pro version, except the Solid tool and the Layout technical drawing tool. SketchUp Is very simple to use, has a clean interface, and comes with a big bank of objects ready to be downloaded. The basic version of the software only has a few functions, but a plethora of plugins are available to meet your needs. However, this software is mainly optimized for 3D architectural modeling. Note that the 3D models you find in the 3D Warehouse are not necessarily printable. You have to make sure that the file matches the prerequisites listed on pages 22-23.

OPENSCAD

This 3D modeling software, known to be the favorite of programmers, uses lines of code as functions. Because its interface is primarily a programmer's text editor, it's next to impossible to understand for someone without a basic knowledge of coding (or computer power users from the 1980s who did everything at the command line). Modeling on OpenSCAD is often basic, but it has the advantage of being usable to create customizable templates for use with other tools such as the Customizer from MakerBot Thingiverse.

3DTIN

This software by LAGOA is TinkerCad's main competitor. It's the "MS Paint" of the 3D world. You can import basic shapes on a plain and relatively simple interface. Just like TinkerCad, this software can only be used with an internet connection which makes it possible to save models in the Cloud. If you like Minecraft, this software is for you.

123D DESIGN

This solid modeling software for beginners can be installed on both Mac and PC and also exists as an app for iPad. It works pretty much like SketchUp, with the usual extrusion, rotation, and drawing functions, but with an interface that's more focused on object design. It has the advantage of creating watertight solids, ready for 3D printing. 123D Design is part of a free software and app package, ranging from 3D photogrammetric scanning (123D Catch) to virtual sculpture, (123D Sculpt), and even to a laser cutting and CNC assistant (123D Make). A downloadable model library is also available, as well as a storage space for online projects, linked to a user community. This software is very useful for creating functional or decorative objects. Its repeat, "loft", and "tweak" functions will enable you to produce complex shapes without difficulty. It is even possible to import existing 3D files or vector drawings to bridge 123D with other softwares like Meshmixer or Illustrator. If you enjoy 123D Design, consider switching to Autodesk Fusion 360, a more complete and professional version.

BLENDER

Blender is very powerful software, capable of generating surfaces of great complexity. it's perfect for the creation of characters and photo-realistic renderings. Its interface includes a lot of functions and it's easy to get lost in its multiple menus. Just like SketchUp, this software is not optimized for 3D printing. You have to take particular care to avoid collision between objects and to open surfaces when exporting the file.

3D SCANNING METHODS

There are several types of 3D scanners. Each has its advantages and limitations, that's why it is important to assess your needs carefully before purchasing a 3D tool. It should also be noted that this is a rapidly evolving technology that might be integrated into some of your portable devices in the years, or even months, after this book has been published.

There are two big families of 3D scanners, those with contact (digital or manual probes) and non-contact (lasers, structured light, etc.). We will only deal with the second category, because the first one is slow, risky (contact with a fragile object) and expensive for use in digital manufacturing. We will also focus on techniques that provide solutions costing less than $10,000 since there is a wide range of architectural 3D scanners, drones, and industrial measurement tools that are definitely too specialized for the purposes of this book.

3D SCANNING BY TRIANGULATION

This type of acquisition tool uses infrared lasers with linear beams to take three-dimensional measurements. The light beam is projected onto the three-dimensional subject, then a digital camera located a few centimeters away from the laser observes the reflections in its field of view. Because we know the angle and position of the laser in relation to the camera lens, we can precisely measure the position of the object, point by point. It is this triangulation information from the laser, the object, and the lens which lends its name to this common technique. By adding a motorized turntable to this triangulation data, we can fully automate the 3D acquisition of an object. One of the most well-known 3D triangulation scanners is the Digitizer from MakerBot, a solution costing approximately $800 that can digitize objects up to 20 cm in diameter and height. With this device, vertical laser beams with filters are located on each side of the lens, to give greater precision. This tool comes with proprietary software that makes it possible to automatically convert a 3D scan to a watertight file that's ready to be printed. The Multiscan feature lets you make multiple scans of the same object and combine them to get to maximum detail. The Matterform and NextEngine brands offer similar solutions to MakerBot.

Name : Matter and Form
By : Matter and Form
Country : USA
Size (mm) : 345 x 210x 354 (foldable)
Max scan volume (mm) : 250 (dia.) x 180
Min scan volume (mm) : 20 x 20 x 20
Resolution (mm) : 0.43
Color : No

Name : Digitizer
By : MakerBot
Country : USA
Size (mm) : 474 x 412 x 204
Max scan volume (mm) : 203 (dia.) x 203
Min scan volume (mm) : 20 x 20 x 20
Resolution (mm) : 0.5
Color : No

MANUAL SCANNER WITH INFRARED LED

These scanners use the same technique as triangulation scanners, the difference being that they are manipulated manually during 3D acquisition. These scanners use an infrared LED, whose invisible light is used to measured the distance between the device and the object. The resulting *point cloud* is transformed into a mesh by the software installed on the computer. This data can be combined with the textures and colors obtained from a camera that's integrated into some scanners. This provides a three-dimensional, multicolored object. Please note that the textures resulting from these devices are often ill-defined and of low resolution because the ambient light has a direct impact on the accuracy of the colors. Among the most popular and affordable solutions are the Microsoft Kinect 1 and the ASUS Xtion Pro Live, both based on PrimeSense technology.

These mainstream motion sensors become very good 3D scanning tools when combined with software like Skanect or ReconstructMe. There are also "all in one" solutions, based on the same technology, such as 3D System's Sense. Also, some infrared scanning tools integrated with touch screen tablets and smartphones (iSense, Structure Sensor) make the technique more mobile than ever by removing the need for the computer. Since Apple bought PrimeSense in November 2013 and therefore owns patents for this technology, we can foresee that some of their future devices will be equipped with it. In the meantime, Microsoft has switched to another technology for its Kinect 2, with more resolution and a wider field of vision.

Name : Sense
By : 3D Systems
Country : USA
Size (mm) : 178 x 129 x 330
Max scan volume (mm) : 3000 x 3000 x 3000
Min scan volume (mm) : 20 x 20 x 20
Resolution (mm) : 0.5
Frequency : 30 FPS
Color : Yes

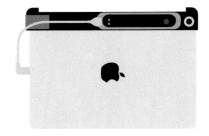

Name : Structure sensor
By : Occipital
Country : USA
Size (mm) : 119.2 x 27.9 x 29
Max scan volume (mm) : 3000 x 3000 x 3000
Min scan volume (mm) : 20 x 20 x 20
Resolution (mm) : 0.5
Frequency : 30 FPS
Color : Yes

STRUCTURED LIGHT SCANNER

Structured light scanners are extremely fast and precise tools. A video projector sweeps the surface of the object while blasting it with patterns, strips, or grids. The projected sequence is observed by a camera which is calculating its deformation. The area of the object where the light is projected can be completely measured in less than a second. Some structured light scanners also include a camera and a white light source, which captures the colors of objects as it passes across them. This enables full color mapping directly on the exported 3D model. Among the most popular structured light 3D scanners are the Artec EVA and the SLS-1 from David Laser Scanner. Fuel 3D, an affordable solution for structured light scanning, was also launched in 2013 by a team from Oxford University.

Name : Eva
By : Artec
Country : Luxembourg
Size (mm) : 262 x 158 x 64
Max scan volume (mm) : 2000 x 1000 x 1000
Min scan volume (mm) : 150 x 150 x 150
Resolution (mm) : 0.5
Frequency : 16 FPS
Color : Yes

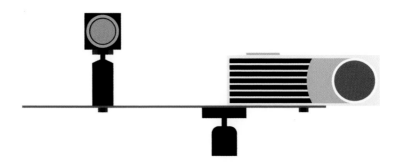

Name : DAVID SLS-2
By : DAVID Vision Systems
Country : Germany
Max scan volume (mm) : 500
Min scan volume (mm) : 60
Resolution (mm) : 0.06
Color : Yes

PHOTOGRAMMETRY
(STEREOSCOPIC METHOD)

3D photogrammetry acquisition solutions are not, strictly speaking, part of the 3D scanning family. The principle is to take multiple photos of an object from every angle and accessible view. The images are then sent to a piece of software using the colors, shadows, and textures as benchmarks in order to re-attach them to a three-dimensional volume. It's the principle of stereoscopy (as in human vision) that's used to calculate the distance between each view. There are systems that can trigger several cameras at once to get a more precise and spontaneous result. These systems generally use SLR cameras, but some prototypes use simple cellphone cameras arranged regularly in a spherical or cylindrical structure. Some software solutions are extremely affordable because they can be used with a simple camera. Among the most popular solutions in this technology is the 123D Catch app by Autodesk. This free and fun tool, available on iPhone, iPad, and Android, is a first step towards accessible 3D. Even if it's difficult to get a watertight model and even if files can take several hours to be calculated, this app adds another string to your bow (and it's free). Photogrammetry is the technique that allows the most exact rendering of colors. This is important if you want to print your model using a polychrome technology, such as the Mcor Iris. 123D Catch is far from being the only solution of this kind available to you. Memento, also a free Autodesk product, is a much more powerful version, ideal for professional uses like museum conservation, architecture, and archeology.

Digital camera
Gives very good results with
Autodesk Memento

iPhone or iPad
For 123D Catch App.

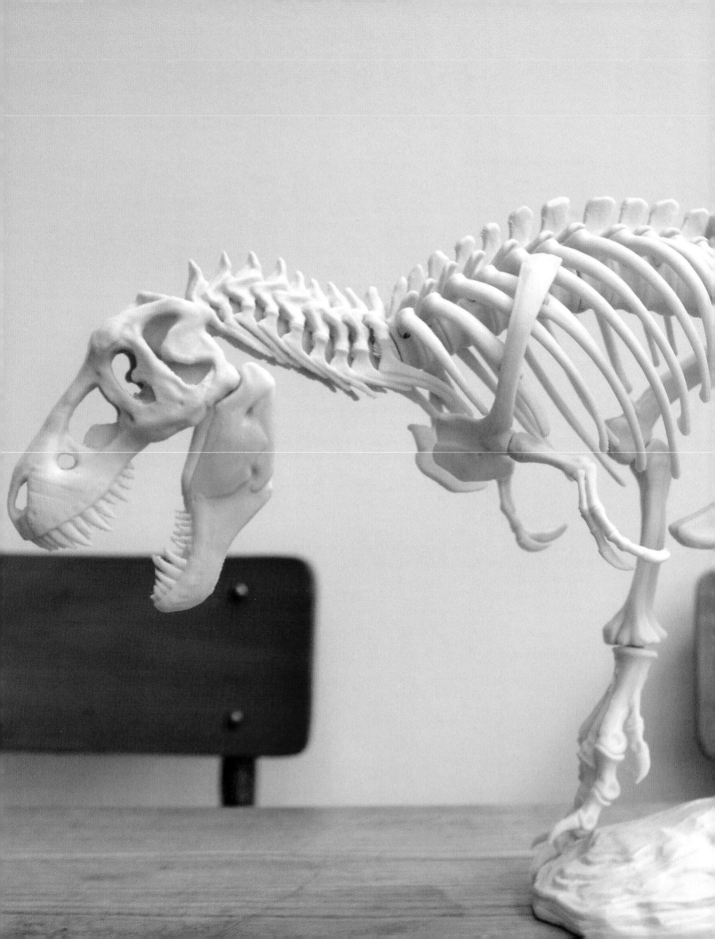

DOWNLOAD

---

*3D PRINTING AT THE
SERVICE OF EDUCATION*

FINDING A 3D FILE ON
THE INTERNET

DOWNLOADING A FILE
FROM THINGIVERSE

TIP: TO OPTIMIZE THE *PRINTING*
OF A SMALL OBJECT

3D PRINTING AT THE SERVICE OF EDUCATION

From theory... to practice: it's the new direction that allows innovation in 3D production in education. Today 3D printing is an invaluable tool for education in technical schools, high schools, universities, and colleges who want to stimulate the creativity of their students. School projects are no longer just drawings or designs on paper. Whether it be to take a design brief from A to Z or to prototype a design, the interest in digital production is undeniable in various environments.

Because 3D printers are so easy to learn, they really favor the development of this teaching method: there are currently 3D printers for all types of projects, and for all levels of learning, from the small Fused Deposition Modeling 3D printer for schoolchildren to the industrial machines used in specialized higher institutes.

Slowly, around the world, schools are equipping themselves at a time when program reforms tend to encourage the learning of computer programming and robotics, from the earliest age. Increasingly in demand, there are not enough engineers and technicians. When faced with a growing demand on the job market: the digital skills of young graduates are valuable in the midst of this new industrial age, and there is no age too young to learn how to be an engineer... or just to be ingenious.

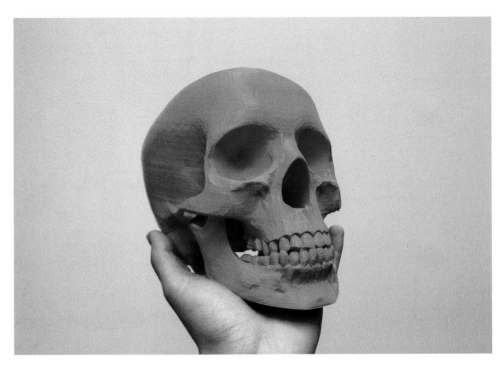

Previous page
T-Rex Skeleton
By: MakerBot
3D printed on: Replicator 2
Source: MB Digital Store
© Samuel N. Bernier

Left page
5-Cylinder Radial Engine
By: MakerBot
3D printed on: Replicator 2
Source: Thingiverse
Thing: 52769
© Samuel N. Bernier

Right page
Human skull
By: le FabShop
3D printed on: Replicator 2
Source: Thingiverse
Thing: 441087
© Le FabShop

FINDING A 3D FILE ON THE INTERNET

There are millions of free 3D files to download on the internet. Using this resource could save you precious time. So before you embark on a long modeling project (an apple, a car, a famous architectural work, etc.), take the time to make sure the work hasn't already been done by someone else. Here are a few platforms where you can find 3D files of all types.

→ THINGIVERSE

Thingiverse is a community created in 2008 by the founders of MakerBot. For a long time it was the most popular in the world of 3D printing but in 2012, the MakerBot's shift from open source to proprietary design divided users. Although shared files on Thingiverse are usually made available under an open license, MakerBot shares some rights to the files published on its platform.

The name Thingiverse comes from the contraction of the English word "Thing" and "Universe." Therefore, Thingiverse is a universe of "things" to produce by 3D printing. There are more than 3 million projects on Thingiverse, ranging from cufflinks to the Eiffel Tower. On Thingiverse, you can create collections, comment on a file's quality, or make a variation or a "remix" of it, or even make a parametric model using the Customizer application.

Thingiverse also holds contests, with fantastic prizes and allows people from around the world to work together on a general interest project, like the Robohand project.

http://www.thingiverse.com/

→ MAKERBOT DIGITAL STORE

If you have children (or grandchildren) and own a 3D MakerBot printer, the digital store is your friend. This addition to the MakerBot ecosystem was launched at the beginning of 2014. Unlike the Thingiverse platform, MakerBot sells original models, in X3G and MAKERBOT proprietary formats. The designs are created by MakerBot internal designers or by studios chosen by the company. Each file is set up and optimized in a way so that the user only has to start printing, without needing to ask questions. For now, the Digital Store houses collections of toys to customize as well as some more complex models such as a T-Rex dinosaur skeleton to put together. It's a promising concept that is currently limited to users of MakerBot.

https://digitalstore.makerbot.com/

→ GRABCAD

This community of engineers has been gaining in popularity for a few years. Its vast high-level file bank is due to a network of users, often professionals, and their free downloads, as well as a very liberal usage policy. GrabCad's competitive advantage is the ability to download projects in original (native) format, rather than simply in STL format. In fact, the website encourages its users to publish their files in different versions to allow anyone to reuse and edit them as they wish. An even more important detail is that GrabCad files are not optimized for additive manufacturing, so often these need to be optimized.

http://grabcad.com/

→ SKETCHUP 3D WAREHOUSE

This gigantic virtual object library was created at the time when SketchUp software still belonged to Google. Widely used by graphic designers during the creation of the 3D version of Google Earth, the Warehouse has an impressive number of files for architecture. If you want to make a model of your IKEA furniture, look no further, because it's surely already there. All files in the 3D Warehouse are free and can be downloaded directly in SketchUp. It's the perfect tool for architects and interior designers looking to speed up the creation of their 3D designs. However, you have to be careful because models created for graphic purposes cannot be printed directly, unless they meet the criteria for 3D printing. Fortunately, you'll have no problem editing them in SketchUp.

http://3dwarehouse.sketchup.com/

→ TURBOSQUID

This website is well-known among video game companies, architects, and others for its photorealistic 3D animation and design. Everything can be found there: people, cars, scenes, furniture, fruits and vegetables... The quality of the visuals is often breathtaking, but don't get too excited. A large number of these files cost money, and not many of them follow the 3D printing rules we described earlier (such as watertightness), because they are optimized for digital (not physical) production. If you plan to purchase a model from TurboSquid, expect to spend some time with your own modeling software to prepare it for production. Don't be surprised if some of them remain unusable despite your efforts.

http://www.turbosquid.com/

→ SKETCHFAB

SketchFab is a website for viewing 3D models with textures and environments. It acts as a support portfolio for several designers from different backgrounds and offers works that are sometimes breathtaking. Some models can be downloaded directly, however, you may have to make a request to the creator via his or her project page.

https://sketchfab.com/

→ YOUMAGINE

This platform specializes in files dedicated to 3D printing and was created by Ultimaker as an alternative to Thingiverse. The community is smaller so there are fewer files, but all of them are trustworthy and original. There are a lot of talented Dutch designers featured there, like Michiel Cornelissen.

https://www.youmagine.com/

→ 123D

In addition to being home to the best free software for beginners, the 123Dapp website hosts a 3D file sharing community, with files that were created using its various applications. You will find many incomplete or basic projects, but among them are rare gems. If you like monsters, be sure to take a tour of the 123D Sculpt+ gallery.

http://www.123dapp.com/gallery

→ PINSHAPE

Pinshape is a platform inspired by social media and has an online store. You'll find both free files and some you have to pay for.

https://pinshape.com/

→ CULTS

The name "Cults" is a reverse anagram of St. Luc (St. Luke), the patron saint of artists and sculptors (and the letters S-T-L are also found in the name). This marketplace created in France allows designers to publish, and also sell their 3D files online. The community is young, promising, and focused on the quality of the projects offered, more than their quantity.

http://fr.cults3d.com/

→ MYMINIFACTORY

Definately one of the best curated website to find cool 3D printable files. MyMiniFactory is filled with projects inspired by the Sci-Fi and Internet culture. Fans of Cosplay and action figures, this is the place for you. You can even sell your designs or be hired for a project.

https://www.myminifactory.com/

DOWNLOADING A FILE FROM THINGIVERSE

CHOOSING THE TYPE OF FILE:

The generic format for 3D printing is STL. If you want to make it simple, make sure that the design that you have chosen can be downloaded in this format. An OBJ format can work easily. If not, prepare for file conversion. Sometimes, a creator will put the native file used to create the model online: SKP (SketchUp), 3DM (Rhinoceros), MB, or MA for Maya, etc. If you use the same software as them, it will be easy for you to edit their model so you can make all the changes you want.

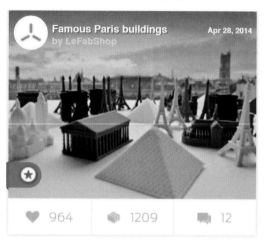

File Name	Downloads	Size
Eiffel_Tower_mini.STL Last updated: 14-04-25	7592	300kb
Pyramide_du_Louvre.STL Last updated: 14-04-25	4272	568kb
Madeleine.STL Last updated: 14-04-25	4539	1mb
Arc_triomphe.stl Last updated: 14-04-25	4823	16mb
sacre_coeur.stl Last updated: 14-04-25	4237	9mb
notredameparis.stl Last updated: 14-04-25	4750	7mb

ALL THE SAME PASSION, BUT NOT ALL THE SAME TECHNIQUE:

An architect doesn't use the same logic for design as an industrial designer who designs for manufacturing. On top of that, different software packages have different tools and options, and each package may have different ways of exporting similar objects. This means that the meshes that they create won't be equivalent. So even if you design the same object in different packages, and export an STL from both of them, both results aren't necessarily suitable for 3D printing (an open mesh, a non-manifold mesh, inadequate resolution, details, or scale, etc.). The best result will come from a software package designed for 3D printing. That said, some enthusiasts have been very successful at creating perfect files using software packages that weren't designed with 3D printing or manufacturing in mind. When you're looking for something to download and print, sometimes the easiest thing to do is read the comments from the community, if there are any, on the file's download page.

THINKING ABOUT THE CONSTRAINTS LINKED TO YOUR 3D PRINTER

Everyone doesn't necessarily have the same 3D printer as you: objects that are too large or projects designed for other technology and materials may not be right for you; an object that prints best in sintered nylon powder may depend on certain flexibility properties that you can't get with PLA fused filament. Remember to make sure you can print the object on your machine by reading the descriptions or comments: and feel free to leave your own feedback and remarks. For the following tip, which explores these constraints, download the miniature Eiffel Tower:

http://www.thingiverse.com/thing:311002

TIP: TO OPTIMIZE THE PRINTING
OF A SMALL OBJECT

PRINTING SEVERAL COPIES

To get good result with fused filament 3D printing, the melted plastic needs to cool down between layers. This can be done in many ways (slower nozzle travel, fan blowing on the extruded plastic... but there is one more trick. Rather than printing small objects one at a time, it's better to print multiple copies of the model on the printing platform. It will give the plastic time to cool down between each layer. Let's print the Eiffel Tower alone using ABS or PLA:

Random chunks of material form on your object when the plastic is too hot. Now, at the same speed and temperature, print the same piece in five copies on your platform.

The plastic stabilizes between each layer, and the nozzle continues to work on the other four objects before coming back to deposit the following layer.Here, the problem is a result of the fact that the Eiffel Tower is very thin at the top: printing a larger monument in one copy would not have the same results, because the layers would be more staggered and placed on top of each other less quickly.

SLOW DOWN FOR SMALL OBJECTS

Most FFF desktop printers allow you to change the extrusion speed and the movement of the extruder. Lower these speeds to prevent smears: this is another way to correct this problem. Be careful, however, to keep the temperature high enough to prevent clogs from forming in the nozzle.

VENTILATING

3D printers equipped with a fan installed on the nozzle allow the layers of plastic to cool off better during printing, especially with the PLA that contracts during cooling. By contrast, ABS must be stabilized at a warm ambient temperature that doesn't fluctuate to avoid warping between the layers.

ADJUSTING THE PLATFORM HEIGHT

For objects that have a low contact surface with the platform, it is better, on printers that allow it, to start the nozzle out closer to the platform: the first layer will adhere better and the piece will have less chance of falling. To adjust the platform height on printers with manual adjustment, you'll need to "cheat" a little when you run through your manufacturer's calibration process (such as using a piece of cardstock instead of paper when measuring the gap). For printers that have an automatic height adjustment, you'll need to check the documentation for your printer to look for an option such as MakerBot's "Z-Axis Offset" or Printrbot's "Z offset."

REPAIRING AND EXPORTING

MINIATURE PARIS

The web is rich in 3D content: there are models of cars, furniture, plants, and even characters from video games... The web probably has what you need. Sometimes these models have a price and sometimes they're free, but the files you can get are rarely ready for 3D printing, unfortunately, because a majority of these files are designed as graphics (not things), whether to create synthesized images, animation, or even architectural layout plans.

The collection of Paris's miniature monuments is an example of some open source architecture files that have been misused with the aim of being printed. Originally created for Google Earth's 3D view, the buildings were downloaded for free on the SketchUp Warehouse platform. The original files, containing hundreds of design errors, such as walls that were too thin, all had to be corrected before being printed on a miniature scale.

This project, completed by le FabShop, had the goal of creating a souvenir collection of Parisian architecture. Each mini-monument, placed inside a small sphere of crystalline plastic, was distributed by a "gumball machine" during the first Maker Faire Paris. The family still needs to be completed with other classics, such as the Moulin Rouge, or the Montparnasse tower, but you can already download the Eiffel Tower, the Arc de Triomphe, Sacre Coeur, Notre Dame, La Madeleine, and the Pyramids of the Louvre at the following address:

http://www.thingiverse.com/thing:311002

Previous page
Paris miniature
3D printed on: Replicator 2
By: le FabShop
Source: le FabShop
© Océane Delain
Source: Thingiverse
Thing: 311002

Left page
Notre-Dame miniature
3D printed on: Replicator 2
By: le FabShop
Source: le FabShop
© Océane Delain
Source: Thingiverse
Thing: 311002

Right page
Eiffel Tower miniature
3D printed on: Replicator 2
By: le FabShop
Source: le FabShop
© le FabShop
Source: Thingiverse
Thing: 311002

REPAIRING AN STL FILE

In order to be used in 3D printing, a file must meet certain criteria (see pp. 22-23).

If you aren't the author of the 3D model and if it wasn't created for digital production, there's a strong possibility that it won't meet one of these criteria.

Here are the most common problems:

→ Reversed normals (Part of the mesh facing in the wrong direction)
→ Open surface
→ Collision between volumes
→ Non-aligned edges
→ Bad file type
→ Wrong scale

There's a chance that you won't find a solution for a model that you didn't create yourself. Fortunately, there is software to solve some of these problems, without having to go back to the model's original source file. Nothing is perfect, because a lot of files are quite simply too corrupt to be recovered, but some distinguish themselves by being remarkably effective. The most effective of them all is MAGICS software, created by the Belgian company, Materialise. This expensive tool caters especially to 3D printing professionals and large companies. In this book, we recommend three free, easy to use software programs:

→ Netfabb Basic
→ Meshmixer
→ Autodesk Print Studio

Each software package may respond differently to the same problem, which is why it is worth it to try several of them to see which will provide the best repair.

For the following exercises, you'll need to download the model of the Eiffel Tower, which has several faces missing. Follow the steps to test the different repair software programs.

https://www.thingiverse.com/thing:440546

REPAIRING A FILE WITH MESHMIXER

Download Meshmixer:

http://www.123dapp.com/meshmixer

The Meshmixer software can be used in several ways. It's one of the rare free software programs that lets you work directly on the mesh of STL files. The "Analysis" function is a major power tool for verifying 3D files.

Click Import, then import your 3D model in Meshmixer, and choose the seventh icon in the bar on the left hand side: "Analysis", which is represented by a sphere dotted with blue triangles. Finally, choose the Inspector tool in the gray box that appears.

If your file contains geometrical problems, colored spheres will appear.

Color code:

● A blue sphere means that there is a hole in the mesh.

● A red sphere points to a non-manifold area.

● A magenta sphere signifies that the components are disconnected from the general mesh.

Left-click on the spheres to correct the problem. If the sphere turns gray, the repair has failed. (Command or Control-right-clicking allows you to choose the area and to repair it manually with other Meshmixer tools.) By clicking "Auto Repair All" in the task bar, all the errors will be fixed at the same time. But be careful, because the software logic may surprise you. Your model may even disappear!

If the automatic repair doesn't work, choose each sphere one at a time to be sure that the repairs have no effect on the geometry of the object. If not, try to perform manual repairs by exploring the different tools in Meshmixer.

In the case of the Eiffel Tower, part of the top disappears completely when it is repaired automatically. In these cases where the "Inspector" fails, there is also the "Make Solid" option in the "Edit" mode (above Analysis). This feature makes your object printable, but it will drastically reduce the quality of its surfaces.

REPAIRING A FILE WITH MESHLAB

Download MeshLab:

http://meshlab.sourceforge.net

MeshLab is an open source, portable, and easy system for the processing and editing of unstructured 3D triangular meshes. The system has been developed to help the processing of the typical not-so-small unstructured models arising in 3D scanning, providing a set of tools for editing, cleaning, healing, inspecting, rendering and converting this kind of meshes.

Open MeshLab and import your broken Eiffel Tower. For this, you just have to click on "File" and "Import mesh". Choose the Eiffel Tower.

The message "Unify Duplicated Vertices" will appears. Check the box, and click "Ok".

To repair our object, go to the "Edit" menu and click on "Fill hole". This will open the following window:

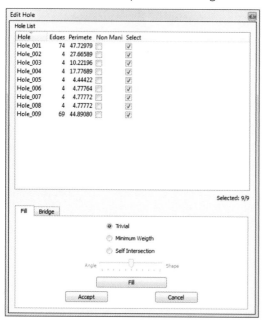

A list of all the hollow faces you can repair is displayed. In this case, check all the boxes, then click on "Trivial" and "Accept". The "Trivial" function aimed at repairing even complicated faces, as we can see in the second section of our object, where some meshes are "broken" inside:

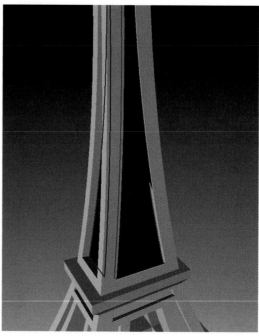

You'll notice that the automatic repair has created some deformities on the long part of the Eiffel Tower. Models that have undergone several repairs will sometimes look rounded or pixelated. However, MeshLab will rarely refuse to repair a file, no matter how complex it may be.

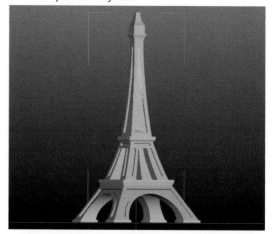

Once your object has been repaired, save a new STL file with the "Export Mesh" function.

REPAIRING A FILE WITH NETFABB BASIC

Download Netfabb Basic:

www.netfabb.com

When you open Netfabb Basic, don't be surprised if a registration page asks you to wait about 10 seconds and accept the terms and conditions before giving you access to the software. This recurring invitation suggests that you upgrade to the Netfabb paid version, with many additional tools for advanced use. Click on "Later" to postpone the request for a license. The Netfabb Basic version may be used for an unlimited time.

The application opens with a gradient gray background. Go to "Project," then "Add part" to upload your 3D model. The object will be displayed in green and red. The red areas symbolize surfaces that have a problem. In the case of the Eiffel Tower, the red indicates the object's "interior" surfaces. So that the model can be printed without any problems, all the visible surfaces must be green.

In the toolbar at the top of the application, you'll see a little red cross called "Repair." Choose this function.

Your object will turn blue and its mesh will be visible. A window will appear at the right with a number of options. Ignore them and select "Automatic repair," at the very bottom. You'll finally have the choice between "Default repair" and "Simple repair." Both of them will work in most cases. Your surfaces are now closed and sealed. Confirm by clicking on "Apply repair" and then "Remove old part".

To save a new STL of your piece, right-click with your mouse on the newly created piece, and then choose "Export part" and specify "as STL".

The automatic repair of the pieces isn't foolproof. You'll see the same deformity on the tower where the tangent polygons were missing.

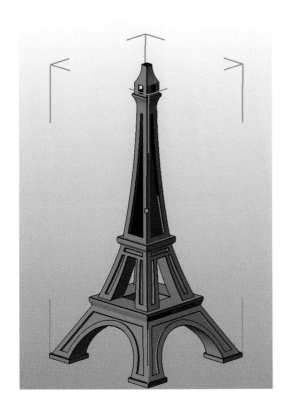

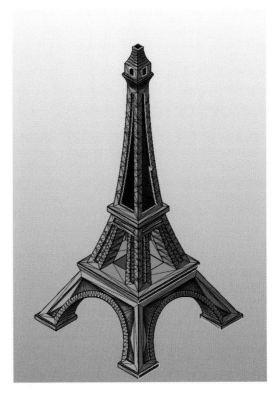

ANATOMY OF A 3D CREATION

Before exporting a file to a 3D printer, you need to know and understand the characteristics of a 3D project created layer by layer.

RAFT

This option creates a plastic base, slightly wider than that of the printed object. This base aims to give more surface adherence between the printed piece and the platform, reducing the risk of separation and buckling. The raft manually separates the piece thanks to a microtexture that weakens contact between the two pieces. When printing with soluble materials, the supports and rafts are removed chemically with Delimonene (for HIPS), water (for PVA), or a warm sodium hydroxide solution (acetal resin). It's better to enable the "Raft" function in the following cases:

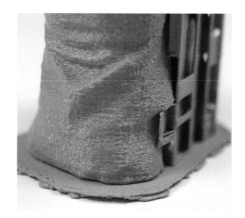

→ Narrow, vertical piece;
→ Multiple pieces on the same platform;
→ Use of a high shrinkage plastic, like ABS;
→ Large pieces (the corners of large pieces have a tendency to bend during printing);
→ Using the "Support" function (the support structure with very thin walls)—the raft gives the platform a solid grip surface;
→ Damaged printing platform (platform surface faults will be on the bottom of the raft and not the bottom of your piece);
→ In case the platform is difficult to level perfectly (the raft has thicker layers, so it is slowly crushed to become parallel to the nozzle's movement axes). With some 3D printers, like the MakerBot mini, the absence of manual leveling on the platform makes the raft mandatory.

SUPPORT (OR SCAFFOLDING)

The support is a detachable structure like the raft, which supports the parts of your objects that are likely to collapse. The support's structure is calculated by your slicing software based on the distances between the bridges, the corners of your print's overhangs, or even layers that float in the void above the platform. Be sure to check that no support will be generated on an inaccessible part of your object, because you'll have to separate it by hand or with tweezers, unless you have used a soluble material (HIPS, PVA).

INFILL

The STL only contains the external walls of your objects: sometimes you will want them to be filled, and sometimes hollow for printing. Expressed as a percentage, the infill will be more or less dense depending on your needs. The higher the percentage, the denser the structure. Usually, these cells

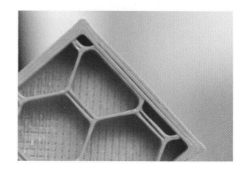

are shaped as hexagons or squares, but many other options may occur (MakerBot even has an infill setting using cat silhouettes). Note that you won't gain much by completely filling your piece: a 20% fill is already quite rigid.

SHELL (OR WALL THICKNESS)

A 3D model's mesh is just a mathematical surface with no thickness: to make an object real, you should give a thickness value to the wall. The "shell" in FFF is the equivalent number of passages that your nozzle will make to draw the outline of your layers. With a 0.4 mm nozzle and a 2-shell setting, your wall thickness will be 0.8 mm: the perfect compromise between speed and sturdiness. Some software, like CURA, will allow you to choose the exact thickness you want for your wall, automatically adjusting the numbers of layers needed.

LAYER HEIGHT - SURFACE CONDITION

With FFF, the objects are generated from layers ranging from 0.1 to 0.3 mm. The object's printed surfaces may appear striated. This staircase effect increases as the surface gets closer to horizontal. So there is a texture that resembles the strata on a topographical map. Surfaces that are just slightly curved will be affected the most. The lower quality the Z resolution (for example, 0.3 mm), the more the layers will be visible. That's why the layers are almost invisible to the naked eye with high resolution 3D printing techniques like DLP. The orientation of the model in relation to the printing surface will not only have an impact on the generation of supports and the condition of the overhang, it will also have a direct effect on the appearance of the printed strata and on the solidity of the piece.

SOLIDITY

You should observe the orientation of the piece's grain closely when printing a functional object, and be mindful of any physical constraints. A narrow prism printed vertically will only have a fraction of the solidity of the same figure printed horizontally. By following the direction of the grain, you can also print objects that are partially flexible, despite using material that is rigid, like PLA or ABS.

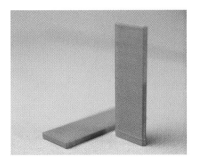

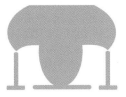

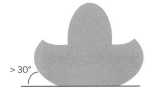

> 30°

EXPORTING FOR 3D PRINTING

Most 3D printers use proprietary or open source "slicing" software, but it is also possible to prepare your prints for slicing using Meshmixer.

First, import your 3D file, and then open the "Print" function.

Print

A new interface will appear. At the top of the new menu, you can choose among a selection of 3D printers (if the 3D printer you are using is not on the list, create your own profile), and you can choose a display color for your model.

Click "Transform" to access more options. The "Fit to build volume" button will adjust your model to the maximum build volume for its current orientation so that it will be as large as possible. The

"Move to platform" button is very important: it assures you that the lowest point of your object is found at the level of your 3D printer's platform. You must always ensure that your print doesn't "float in space" (otherwise, useless supports will be printed) or your object won't be situated under the printing platform. Once you are satisfied, you can can click "Done", then use "Add Supports" to create a fractal support structure that will save you time and material.

The areas to be supported will turn red, outlined in blue: you can modify these areas in the "Overhangs" toolbox, which may be accessed by clicking the tool icon to the right of the Supports button. If you aren't satisfied with the orientation of your piece, an "Optimize Orientation" button, at the bottom,

will calculate the orientation direction that requires the least support (but not necessarily best for the aesthetic quality of the piece). Once you open the "Support Generator," it will give you access to the structure qualities: diameter of reinforcement pins on the base or on the top, etc. We recommend always having structures that are more than 2 mm in diameter, so you won't have thin structures that may cause problems with some 3D printers.

Visualize your support by using the "Support All Overhangs" button and start over by deleting it with "Remove All Supports." Once you are satisfied, one click on "Done" and you'll be returned to the Meshmixer "Print" interface... This time, with a support that is ready to be exported, you can choose "Export" at the bottom of your interface, to get an STL of your piece and its support, or choose "Send To," which will open automatically your file in the proper software for your 3D printer, if supported by Meshmixer (MakerBot, Dremel and Type A printers are on that list)

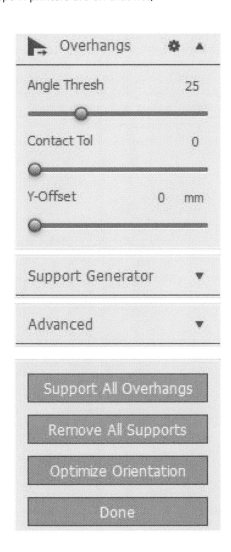

Remember: an STL file isn't a format understood by 3D printers, but only by their software. You'll need to go to your printer software to slice and convert your model into machine language (Gcode, X3G, Zcode...).

CUT AND COPY

PROJECTS IN KIT

Alexandre Kournwsky is a young, ambitious industrial designer. The year he obtained his diploma from ENSCI – Les Ateliers, a renowned design school in Paris, he conceived Meïso, a relaxation center dedicated entirely to flotation and meditation. To carry out his project, the young entrepreneur had to first build a flotation capsule prototype, a sort of cocoon partially filled with water enriched with magnesium salt. He first considered thermoforming a large plastic sheet on a mold made of CNC-milled medium-density fiberboard (MDF) panels, but the tooling and manufacturing costs were exorbitant. So, he devised a technique using of a small 3D printer to make the mold brick by brick. The 3D model of the original mold was thus split into 450 parts, which he printed one by one on his MakerBot Replicator 2. It took three months to make all these bricks. Once he finished, the blocks were glued together, like a monumental 3D puzzle. Assisted by his friend and partner, Maïté, he used a mixture made from hot glue sticks as mortar for the construction of the large PLA igloo. Once it was assembled, several days of finishing with car mastic and sandpaper were essential before obtaining the smooth and even surface of the mold shown on the facing page.

What is probably the largest mold in history to be made on a personal 3D printer was used to make a first prototype of the cocoon in fiberglass. The first brick of the Meïso company.

Alexandre Kournwsky's project proves that 3D printing has no limits for whoever offers their time to make their dreams come true.

Previous page
Building the mold
By: Alexandre Kournwsky
3D printed on: Replicator 2
Source: Meïso
© Alexandre Kournwsky

Left page
Meïso's mold finished
By: Alexandre Kournwsky
3D printed on: Replicator 2
Source: Meïso
© Alexandre Kournwsky

Right page
Architecture model
By: Alexandre Kournwsky
3D printed on: Replicator 2
Source: Meïso
© Samuel N. Bernier

SPLIT AN OBJECT WITH NETFABB

Sometimes the object we want to make is too wide or too tall for our 3D printer. In this case, one trick is to split the volume into several components to be made separately, then assembled. When you create the 3D file, it's best you make these cuts directly in the software as you make the design. If you can't do that, it's possible to split large 3D objects using the Netfabb Basic and Meshmixer software. This trick also works when you want to print a geometrically complex part without support material or when you want to create a flat surface on an object whose base is not. Here's how to do it:

Before beginning this exercise, download the Eiffel Tower from user B9Creations on Thingiverse:

http://www.thingiverse.com/thing:22051

This tutorial shows how to change the scale and split a file found on the Internet in order to make it into a format exceeding the size of the 3D printer.

After retrieving the Eiffel Tower model, open it in Netfabb Basic. The original object is 120.95 mm tall, which is too small to be printed in detail on most FDM/FFF 3D printers. We'll multiply its size by about five times to obtain a 60 cm tall tower, which is four times the maximum height of a MakerBot Replicator 2.

In the toolbar, select the "Scale" icon, represented by a gray sphere with two arrows, one vertical, the other horizontal. This opens the Scale Parts window. In the "Parameters" section of the window, go to the line indicating "Target size" and change the Z value to 600 (the units are in millimeters). Then, click "Scale." The model will automatically get bigger giving the impression that you've zoomed into the center of the part. Spin the scroll wheel

on your mouse towards you for an overall view of the part again.

To be precise during the next steps, it's best you replace the isometric view of the part with a simple front view. For this, select the "Front" icon in the toolbar.

You can now begin to divide the Tower. On the bottom right of the interface, you will find a gray window called "Cuts." Set the Z-axis slider to 279 mm and leave it in the "Cut all parts" mode. Click on the "Execute cut" button. A new window will open. Click on "Cut" again.

You will see a new component appear in the "Parts" window on the top left of the screen.

Now repeat the cut function at the following heights: 140, 0, -29, -32, -229.

You should then have seven groups, but three of them are still too big for a Replicator: the first level, the feet of the Tower and its bases. Click each of these two volumes while holding down the shift key. The area should be colored green while the rest remains gray.

In the "Cut" function, move the X slider to any value and then back to zero, then switch the cut mode to "Cut only selected parts.". Execute the cuts. You can repeat the step on Y with the bases and feet selected.

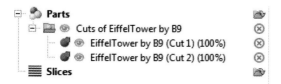

Your model now contains twelve volumes, but you only need to save seven of them under the STL format, as several of them are symmetrical and identical.

Upon exporting, accept the "Optimize" option suggested by Netfabb. This function will automatically resolve any conflicts generated during the previous step.

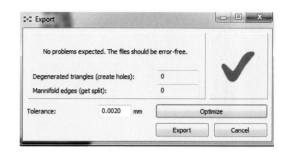

© Le FabShop

The parts, once printed, can be assembled with Epoxy glue.

You can download the split files on Thingiverse to compare your result:

http://www.thingiverse.com/thing:399769

The parts available on Thingiverse are positioned so as to optimize printing times and to eliminate the need of support materials or rafts.

The technique shown here to create a 60 cm Eiffel Tower can be used for a multitude of other applications. It makes the creation of objects in various colors and materials easier, allows you to optimize the printing direction and makes the production of objects sized much larger than the volume of the machine.

© Le FabShop

SPLIT AN OBJECT WITH MESHMIXER

When you use a FDM/FFF 3D printer, uneven and organic volumes are often problematic, as their lack of a flat surface requires support structures. In certain cases, when it doesn't affect the design of the object, you can artificially create a flat surface by creating a material removal in relation to a surface. Here's how to do it with Meshmixer:

Open Meshmixer and select "Import Sphere" in the central window or in the "File" tab.

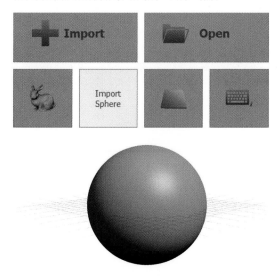

A generic sphere will automatically import. In the toolbar on the left, choose the "Edit" function, represented by a gray sphere strewn with black triangles. Select the "Plane Cut" function." A netting will appear, transforming half of your sphere into a phantom area. If you need to change the orientation of the plane cut, just drag your cursor

Edit

onto the green rotation axis and slowly spin the cutting plane using the ruler of the dial that appears. Drag your mouse to the ruler as you rotate it to

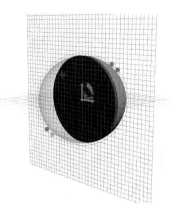

snap to 5 degree increments. To move the plane up and down, select the small blue arrow of the vertical axis (Z), then drag the plane to the desired

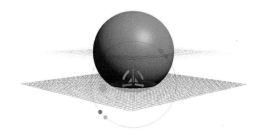

height. Make sure the large blue arrow is oriented towards the face you want to remove. Change the orientation by clicking on it.

When you make the cut, the hole will automatically close up as long as the perimeter of the cutting area is even and the "Fill" mode of the "Plane Cut" window is set on "DelRefine"or "Delaunay". (Changing the cutting style of "Cut" to "Slice Groups" allows you to split the mesh into different shells. To be able to select them separately, you will have to open the "Object Browser" and use the "Separate Shells" tool in "Edit".) Click "Accept" in the gray window to confirm your choice.

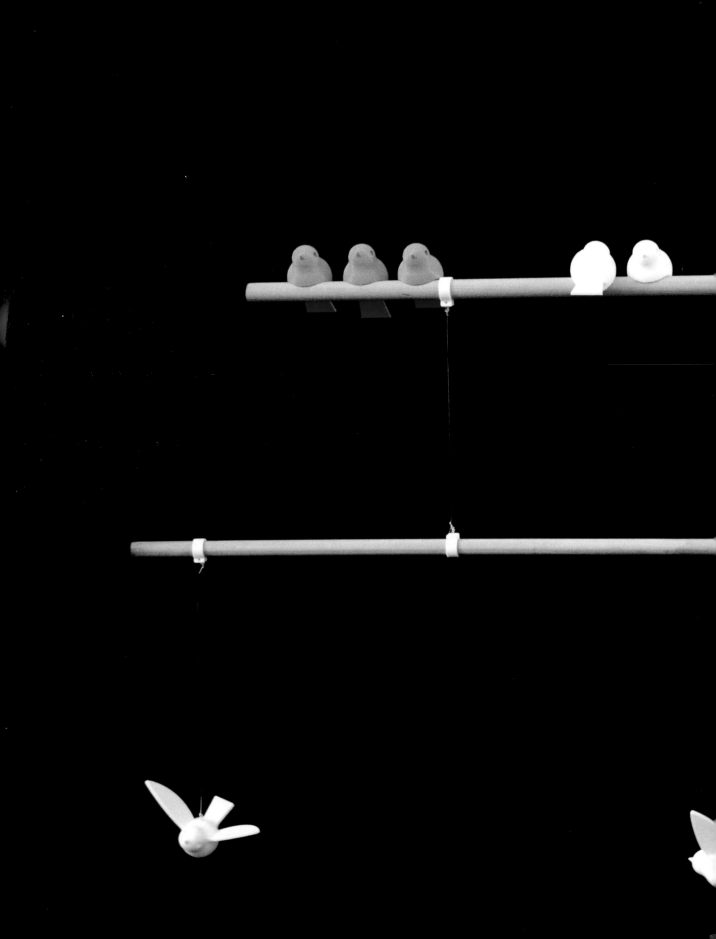

3D MODELING

MODEL A HANDMADE GIFT

You've probably given a Play-Doh sculpture or a macaroni picture as a Mother's or Father's Day gift. These objects, although aesthetically questionable, remain throughout the years in attics and memory boxes, as we hold unique and handmade objects with particular affection.

The "Mobile for Joseph" is an example of a "gift 2.0". In December 2013, the financial Director of LeFabShop became a father for the first time. So, his team of designers consulted with each other to figure out what gift to give him to celebrate the arrival of his newborn. The same day, a small mobile balanced by birds and clouds was created in the 123D Design software.

The initiative was documented then shared on the Instructables and Thingiverse websites so that the design could benefit hundreds of other infants and parents.

It takes a considerable number of hours of training to reach the level where you can produce an original idea in 3D. The mobile exercise is a first step in learning modeling for three-dimensional printing.

You can check out the original project on :

http://www.instructables.com/id/Birds-Mobile/

Previous page
Joseph's mobile
By: Samuel N. Bernier
3D printed on: Replicator 2
Source: le FabShop
© Tatiana Reinhard
Source: Thingiverse
Thing: 269088

Left page
Joseph's mobile (birds)
By: Samuel N. Bernier
3D printed on: Replicator 2
Source: le FabShop
© Tatiana Reinhard

Right page
Joseph's mobile (cloud)
By: Samuel N. Bernier
3D printed on: Replicator 2
Source: le FabShop
© Thomas Thibault

MODELING FOR 3D PRINTING

Several parameters exist that must be taken into account in order to create a "good" file compatible for 3D printing. Each case is unique, but there are some basic rules to follow in order to avoid unwelcome surprises at production time.

The object created on your software must have all of its surfaces closed in order to create a geometry whose interior and exterior are clearly defined. The term "watertight" is used to define this property. Problems with watertightness are very rare if you only use "solid" modeling functions on software like SolidWorks, 123D Design, Catia, or Inventor. On the other hand, you might make some mistakes if you're a beginner on Rhino, SketchUp, or even Blender. A simple method to check the volume consists of visually examining all the seams of the 3D object and fusing the badly "sewn" surfaces. Some 3D software or their plugins simply won't let you export a STL file if the geometry isn't perfectly defined. Sometimes surfaces are completely reversed when imported into another software. In these cases, Netfabb Basic offers an automatic repair solution (see the exercise on page 45).

It's important to know the scale on which the object will be made before rushing into modeling. Not every machine has the same printing volume. Determining the size of the printed object allows you to choose the machine to adopt or even to plan an assembly method in case you need to produce it in multiple parts. (see the exercise on page 54) This is particularly important for people working on an architectural scale. For example, a 3D file of a house modeled at its actual dimensions will probably have a problem of unprintably small thicknesses when it comes to the windows, handles and other details that, once scaled-down, will no longer be able to be printed. For example, a 10 cm wall at a 1:100 scale will only have a 0.1 mm thickness. For peace of mind, allow a minimum of 0.8 mm of wall thickness for most 3D printing techniques. Increase this thickness if you want to refine your piece or apply another finishing treatment (see page 134). The 3D printing service Shapeways, for example, asks for a minimal thicknesses of 2 mm for client files using powder sintering. In every case, it's best to take precautions and avoid going below the recommended thicknesses.

In the case of FFF or FDM 3D printers, it's interesting to take into account the opening of the extruder nozzle. Most "personal" or "office" 3D printers on the market are equipped with brass nozzles that have a circular hole 0.4 mm in diameter (regardless of whether the filament used is 1.75 or 3 mm). A line of material has, as a result, a width of 0.4 mm. If you print a hollow cylinder, its minimal wall will be 0.4 mm, like the case of the famous "Stretchlet" bracelet by designer Emmett:

http://www.thingiverse.com/thing:13505

Note that the scale of a 3D file can be changed at any moment and as often as necessary, whether it be directly in the modeling software, using the 3D printer's software (MakerBot Software, Slicer, Cura...) or by means of a middleware (Netfabb, Meshmixer...).

MODELING USING "PRIMITIVES" ON 123D DESIGN

Download and install 123D Design:

http://www.123dapp.com/design

The 123D Design software from Autodesk is a great tool for creating simple 3D models. In addition, the use of solid modeling principles almost automatically insures the feasibility of the parts by 3D printing, as long as the bodies remain fused. Directly connected to other software of the suite, 123D Make and Meshmixer, 123D Design is the heart of a comprehensive ecosystem for digital manufacturing. This exercise will show you how to use primitive volumes to create a small bird in fused deposition modeling.

Open a new project in 123D Design.

In the toolbar on the top of the interface, select "Primitives," represented by a half circle hidden behind a cube.

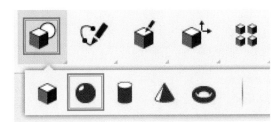

Choose the sphere. Position this sphere at the origin of the grid (position 0,0) and give it a radius value of 14 mm.

Add a second sphere at 15 mm on the X axis of the grid then give it a radius of 8 mm.

Left click on this sphere, then select the "Move" tool among the options of the window that just appeared on the bottom of the screen.

Drag the up arrow and move the small sphere up to 12 mm. It should come in contact with the large sphere, representing the bird's body.

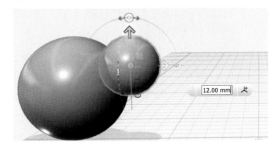

Use the orientation cube (top right) to set your work area to Front view. Now import a cone, also one of the "Primitives" options, and place it at 25 mm from the large sphere, following the grid. Give it the following values: Radius = 3, Height = 6.

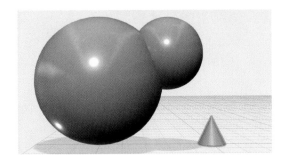

Then, select the cone by left-clicking. With the "Move" tool's rotate icon (just above the arrow), spin the volume 90 degrees clockwise, then slide it vertically to 17 mm.

Then, import a cylinder 15 mm from the large sphere, still following the floor grid. Give it a radius of 2 while keeping its height.

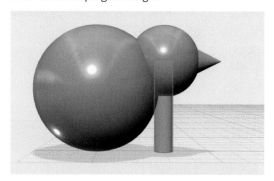

Spin it horizontally with the "Move" function, then slide it vertically 11 mm from its original position so that the cylinder forms the bird's eye.

Now use the orientation block on the top left to place your work area in plan view (Top).

Insert the center of a primitive box at -20 mm in relation to the origin while giving it dimensions of 15 mm in length, 30 mm in width, and 2 mm in height. Go back to front view and move the box vertically to 10 mm so that it becomes the bird's tail.

In the toolbar on top, select "Combine," represented by one cube hiding a second. Three options will appear. Choose the first (Merge), symbolized by a cube fused to a circle.

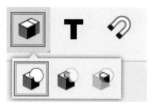

Once the function is turned on, click on the bird's tail, body, head, and beak. Then press the "Enter" key on your keyboard. The four volumes will now form just one.

Now, still with the "Combine" tool, use the "Subtract" function, second on the list. This time, first select the bird's body in "Target Solid," then the cylinder of the eye second in "Source Solid." When you confirm the action with the "Enter" key, a hole should have replaced the cylinder.

All that's left is to create the neck and the curves of the tail. For this, we'll use the "Fillet" function, represented by a cube with rounded corners, located in the "Modify" tool. You can also use the shortcut key (E).

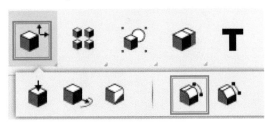

Select the ring between the head and the body, then give the function a radius of 10 mm. Keep the "Tangent Chain" option checked off. Your bird now has a neck.

Switch to Top view, then Fillet again, selecting the small edges on each side where the tail meets the body. Give them a "fillet radius" of 30 mm.

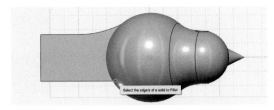

Use the filet function one last time for the semi-circular edges above and below the tail. Give the filet a value of 5.

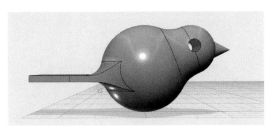

Your bird is modeled. All that's left is to create a form on which it can be attached. In the case of a "Mobile for Joseph," the birds hook onto a wooden dowel 15 mm in diameter. We can repeat the technique we used to make the cavity for the eye. Unleash your imagination and then draw other forms that will let you hang it on other surfaces, like from the top your computer or even the tip of a pencil. Save your new friend on your 123D account or on your computer, but don't forget to export a STL version for 3D printing!

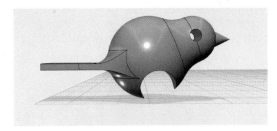

In the "slicing" software of your 3D printer, a printing trick for the bird is to print the tail first so that it can be made without using material

support. Thanks to its shape, it can also be built completely hollow with an "infill" filling of 0%. However, don't hesitate to use the "Raft" function to stabilize your bird during printing. Otherwise, it may fall before being finished. To obtain a better resolution on small details, such as the bird's beak, we recommend you throw several copies of the model on the same tray. This will allow the melted plastic to cool properly in between each layer.

DIFFERENT APPLICATIONS USING "PRIMITIVES"

To create the mobile's clouds, import 3 spheres of different diameters (here 20 mm, 30 mm, and 40 mm).

Cluster 5 large spheres together by using copy and paste shortcuts and placing them randomly. Make sure not to leave any gaps in between.

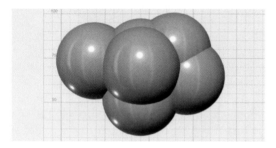

Create 5 medium-sizes spheres and attach them to the others. If you want, use a third size.

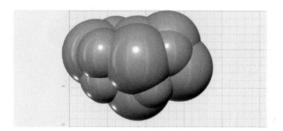

With the "Merge" function, fuse all the spheres into a single volume.

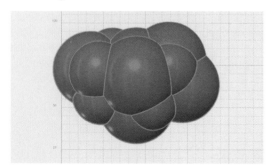

Then, create a large box exceeding the clouds in length and width. Place it so that it runs across the lower portion of the clouds.

With the "Subtract" function of the "Combine" tool, subtract the box from the clouds. To make the model softer, use the filet function to round the edges at the base of the clouds. If needed, add some holes to hang them from the mobile.

When you're satisfied with the shape of your clouds, send it directly to the 3D printer with the "3D Print" option located on the main menu.

Your file will open in Meshmixer and all that's left for you to do is to orient it for printing.

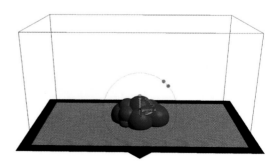

MODEL ON 123D DESIGN USING SKETCH FUNCTIONS

Although using primitive volumes is sometimes very convenient, it's often best to model using drawings to which you can apply extrusion, revolution and subtraction functions.

This small example shows how to create an elastic band used to hang the clouds from the wooden dowel mobile.

To begin, draw a circle 12 mm in diameter centered at the origin by using the sketch function "Sketch Circle."

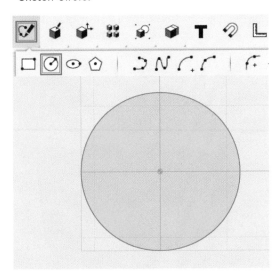

Using the "Polyline" tool, create a rectangle 2 mm large starting at 5 mm from the center and passing the circle another 5 mm.

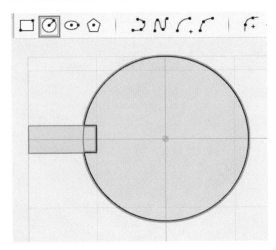

Use the "Trim" tool to erase the lines inside the outline and at the left end of the rectangle.

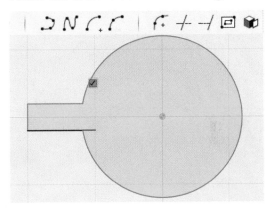

With the "Sketch Filet" tool, create some curves with a 1 mm radius in between the circular and vertical sections.With the "Offset" tool, select the outline to create a 0.8 mm outer gap.

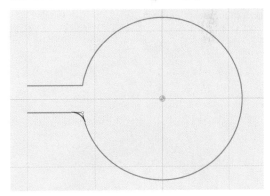

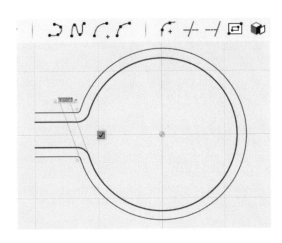

Go back to the "Polyline' took to completely close the sketch. It should automatically be filled in light blue. This is a sign that the drawing can be used for a function.

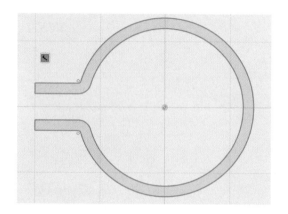

In the "Construct" tool, use the "Extrude" function to give your drawing some volume by selecting it and giving it a value of 5 mm.

Now place your work plane into the Right view and use the "Sketch Circle" tool to create a circle at the center of the flat part. You must select a smooth surface to transfer the sketch plan there. Use the "Polyline" tool to find the center of the rectangle. Give the circle a diameter of 1.5 mm.

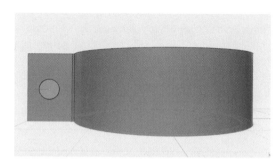

Using the "Press Pull" function, from the "Modify" tool, make a hole through the part. You can also use the P key as a shortcut. A red cylinder should appear as a sign of material removal.

You've made a technical part using drawings on 123D Design!

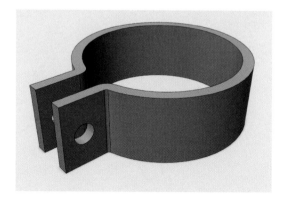

TIP: MAKING FLEXIBLE PARTS

FLEXIBLE PARTS

Special filaments such as Filaflex and NinjaFlex can be used to create flexible objects with a rubbery feel. However, an object can achieve flexibility through design techniques such as incorporating hollow shapes and thin walls.

The PLA bracelet linked here is a good example of a shape made flexible with simple design rules such as open contours and minimum thickness.

www.thingiverse.com/thing:445476

TELESCOPIC PARTS

With 3D printers, it is easy to create objects nested inside each other like Russian dolls. The bridges exercise on page 82 is a good example. If we superimpose slightly conical parts in one file, we can get assemblies with telescopic movements!

This has been done in the following file:

www.thingiverse.com/thing:445481

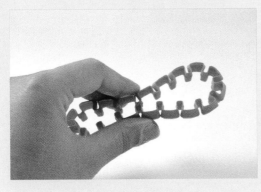

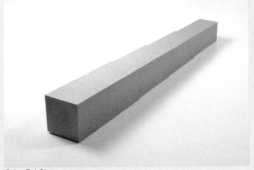

© Le FabShop

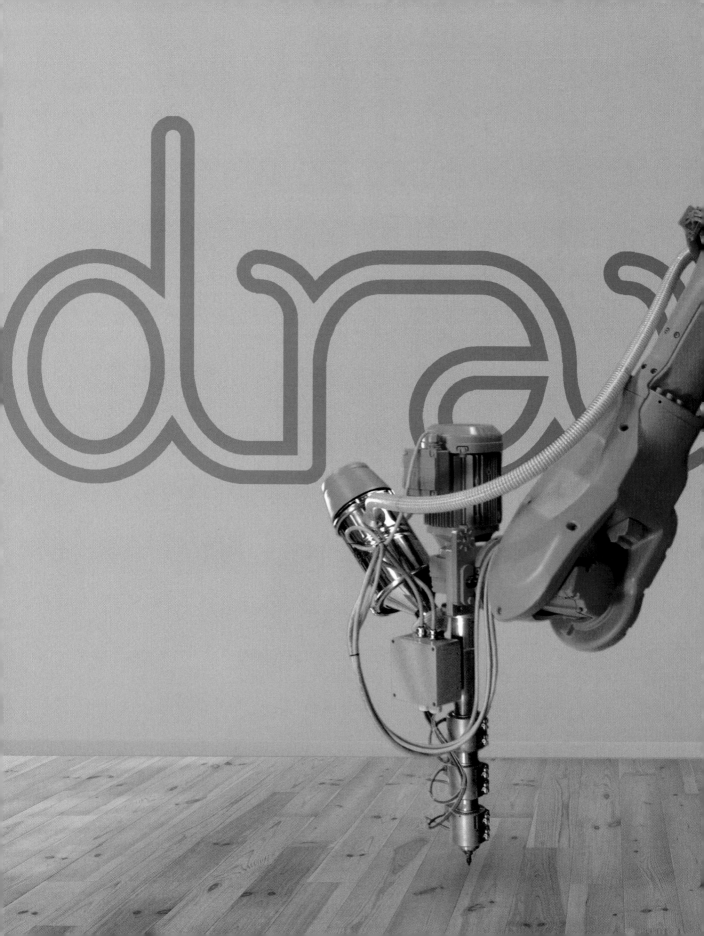

DRAW

WE ARE ALL DESIGNERS

MODELING A COOKIE CUTTER FROM
AN SVG FILE WITH 123D DESIGN

WE ARE ALL DESIGNERS

The democratization of digital manufacturing enables more people to prototype their ideas. What kid has not dreamed of seeing his drawing coming out of the paper and taking form? The layer-by-layer process is just a superposition of stacked 2D drawings, and even if you have no modeling skills, you can design a three-dimensional shape quickly from an extruded drawing by giving it depth.

A childhood dream was brought to life in a project called Drawn, using a gigantic home-made 3D printer named Galatea. The machine can create a functional chair with minimal human intervention, in less than two hour. Galatea is actually a re-purposed assembly robot from the car industry. The tool on its end was replaced by a plastic extruder laying thick layers of ABS on a large heated bed.

In 2015, the company completed a Kickstarter campaign to launch a furniture collection, but the main mission is to spread the concept in different countries to enable customers to design their own objects and get them produced locally by the closest Galatea. Virtually, any sketch of a chair's profile could be transformed into an actual piece of furniture with minimum effort. The machine is better at producing open profiles because filling volumes with melted plastic would cause retraction and cooling problems. Therefore, most of Drawn's designs are extruded silhouettes, perfect for fast and material-saving 3D prints.

Previous page
Galatea 3D printer
By: Drawn
Source: http://drawn.fr/
© Drawn

Left page
C2 Rounded chair
By: Drawn
3D printed on: Galatea
Source: http://drawn.fr/
© Drawn

Right page
Dawn kid rocking stool
By: Drawn
3D printed on: Galatea
Source: http://drawn.fr/
© Drawn

MODELING A COOKIE CUTTER FROM AN SVG FILE WITH 123D DESIGN

An SVG file is a vector format: you can create one in Adobe Illustrator or one of its free competitors, such as Inkscape. We'll use it as a base to create a cookie cutter. For this cookie cutter, we chose a "mustache" shape, which you can download from thingiverse.com (**http://www.thingiverse.com/thing:311668**) under the file name *MustacheLine_SVG.svg*.

If you have another idea, draw your own design in Illustrator or Inkscape. You can also find thousands of free vector files online, such as logos, icons, and drawings. An outline is enough, the only thing to keep in mind is scale. Work in a document using millimeters as units and create a design the size of your future cookies: the original mustache drawing is 70 mm long, so about 2.75 inches.

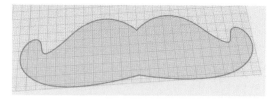

Open the SVG file in 123D Design: click the drop down menu and chose "Import SVG » As Sketch." Select your design: a scaled version of it will appear into your workspace. We then have to extrude the drawing to make an object: select the "Extrude" tool in "Construct" and click on the outline of the sketch. The height will define the depth of the cookie cutter: 8 mm should be enough. Press "enter" or left-click to validate.

Let's move on to discover a new tool: "Modify » Shell".

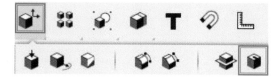

After selecting this option, choose the top face of the model. The volume will open, leaving only a thick outline. You can choose this thickness by pulling on the white arrow or using the gray box at the bottom of the screen.

In this case, the thickness of the "shell" should be 0.8 mm, equivalent to two layers of nozzle passage for most FDM/FFF printers. The shell should be thin enough to cut through dough easily. Make sure to create the shell toward the outside of the sketch to respect the original proportions.

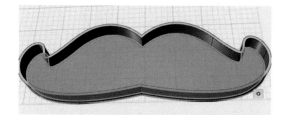

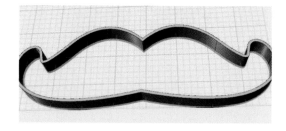

Then, completely open the volume by removing the bottom of the shell. For that, select the imported sketch again and, with the "Construct » Extrude" tool, pull the white arrow through the bottom until

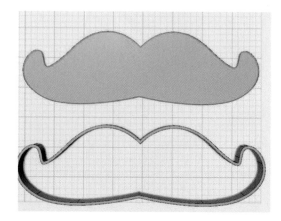

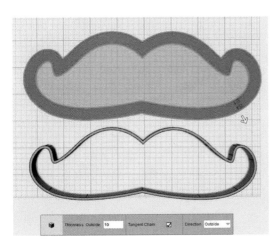

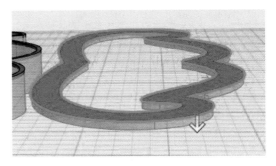

You can now export your cookie cutter as an STL or send it to Meshmixer.

the zone turns red. Once you accept, only the 3D outline of your drawing should remain. Do not delete the sketch base, we will need it to create a small reinforcement for our cookie cutter so that it doesn't lose its shape. You can move the first 3D model to the side for later use.

Make sure to fill your FDM/FFF 3D printer with food safe filament (FDA approved). You will be able to cook your custom cookies in less than an hour.

Select the sketch and extrude it as a New Solid-with a thickness of 1 mm. Select it with the shell function, and choose 2 mm toward the outside to create a thick rim. As you did for the first part, subtract the inside material using the extrusion tool, using the original sketch.

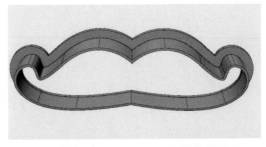

You can now bring back the parts together using the"Snap" tool.

Select the top face of the rim and the thin thin underside of the shell. The first volume selected should then be attracted by the second one, overlapping each other perfectly. Use the "Combine » Merge" function to turn both volumes into a single mesh.

You can now preheat the oven!

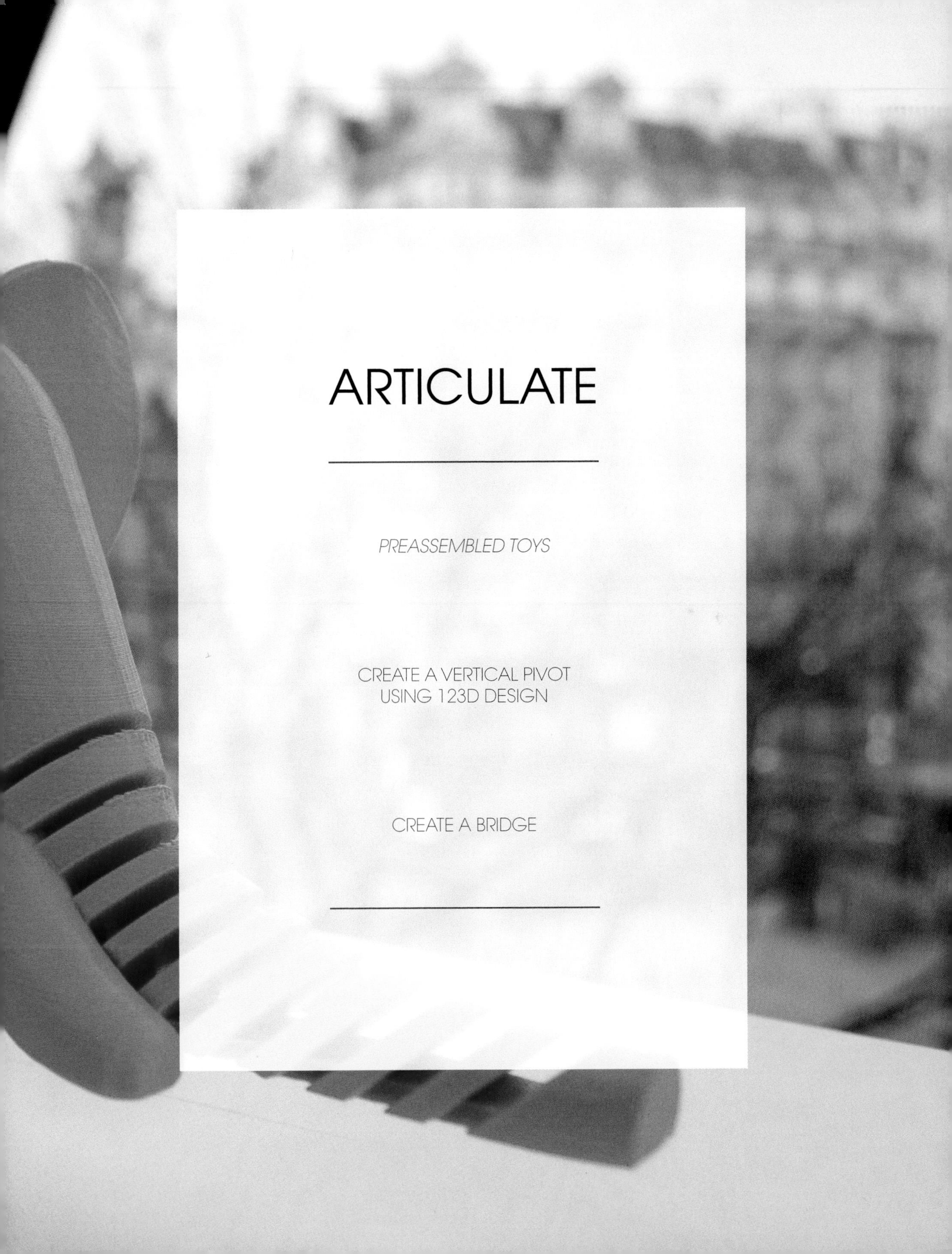

ARTICULATE

PREASSEMBLED TOYS

CREATE A VERTICAL PIVOT
USING 123D DESIGN

CREATE A BRIDGE

PREASSEMBLED TOYS

One of additive manufacturing's most significant advantages is its ability to produce objects with components that are fully-assembled and movable straight out of the machine.

Traditionally, injection-molded plastic objects are made following strict criteria of uniform wall thickness, release angles, and assembly tolerances. 3D printing, thanks to its method of building up successive layers, enables the creation of variable thicknesses, with no effect on the surface quality. As there is no need to mold, it also gets rid of the need to create *draft angles* to remove the finished object from the mold more easily. Assembly tolerances are still necessary, but in some cases, different parts of an assembly may be manufactured and joined simultaneously.

The elephant, robot, and car on these pages were made on a MakerBot Replicator without any support material. The elephant's legs, head and trunk, like the wheels of the car, and the robot's joints are movable straight out of the machine. Such objects can't be created in one go with conventional industrial processes.

The following exercises will teach you how to create moving parts for manufacturing using FDM/FFF by creating pivot axes and bridges. These tips will eventually help you to imagine more complex articulated objects such as Makey the robot.

Previous page
Articulated elephant toy
By: Samuel N. Bernier
3D printed on:Replicator 2
Source: Thingiverse
Thing: 257911
© Thomas Thibault

Left page
Makey the robot
By: Samuel N. Bernier
3D printed on:Replicator 2
Source: Thingiverse
Thing: 331035
© le FabShop

Right page
FabShop Mobile
By: Samuel N. Bernier
3D printed on:Replicator 2
Source: Thingiverse
Thing: 359008
© Tatiana Reinhard

CREATE A VERTICAL PIVOT USING 123D DESIGN

In a new 123D Design project, import a cube type using the "Primitive Box" function. Place its bottom left corner at the start of the grid.

Then select the "Polyline" located in the "Sketch" category, then draw a profile starting from the face of the box with the following dimensions:

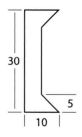

Next, use the "Sketch » Offset" tool to create a +0.5 offset of the first sketch. Delete the box.

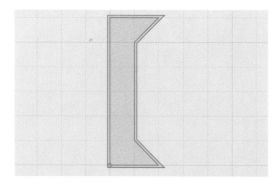

From this offset outline, create a second profile (shown in light blue in the next image) using the sketching tools. Make sure the top and bottom of the second profile line up with the first one.

Feel free to toggle the "Snap" function, found at the bottom right, to help you use the grid or the sketch as a guide. Use the "Sketch » Trim" tool to remove lines you don't need. Then create a 1mm wide rectangle, enclosed within the second sketch.

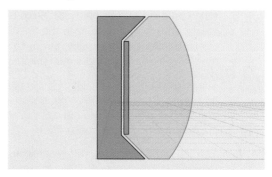

Using the "Revolve," function, found in the "Construct" tool, select the first profile and the newly drawn rectangle. For the axis select the whole left edge of the first sketch.

Revolve it 360 degrees:

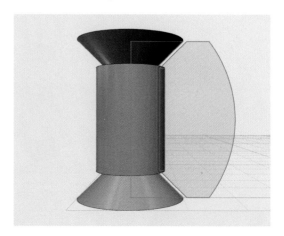

There must be a minimum gap of 0.5 mm separating the two parts to ensure that they move relative to each other after printing. Extrude the second profile giving it 1 mm on each side (2mm in total)

using the "Construct » Extrude" function. You may have to move it 1mm to center it with respect to the revolved shape.

Then choose the "Circular Pattern" function in the "Pattern" toolbox.

Select the shape you just extruded as the "Solid" and the circular edge of the central part as the "Axis". Give the function a value of 12. Now the fan blade should be duplicated 12 times:

Join the 12 fan blades to the cylinder crossing them by using the "Combine » Merge" function.

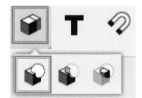

Using the "Shell" function in the "Modify" toolbox, select the top of the axis to create a cavity.

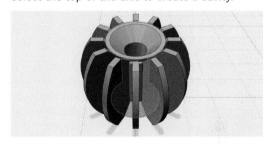

Give it a thickness of 0.8 mm. Finally, add a 2 mm chamfer (shortcut key C) at the base of the central axis. This will ensure the first layer of melted plastic (if FDM/FFF is used) dont spread and merge the two parts together.

Note that, for this model, the angles and spaces between the parts are designed to print the object using a single material filament without support. Start 3D printing your fan with the closed part of it down. Use the "Raft" function to stabilize the elements while production is taking place. If your

FDM/FFF machine is calibrated correctly, the raft should detach without any problems and the object will rotate on itself.

It's best to avoid changing the scale of preassembled objects such as this one, because the tolerances that allow play between the components could become too tight or too loose.

CREATE A BRIDGE

In fused filament 3D printing, pivot assemblies are not limited to vertical axes. It is possible to create hinge and chain effects due to the FDM/FFF's ability to connect the horizontal distance between two points without support material. This feature is called "Bridging".

It is thanks to this trick that the axes of the elephant's legs and head (see page 77) could be created. It's also thanks to this that the Makey robot's legs are articulated.

To see the principle up close and live, here's a simple model that will make you a "Bridging" guru.

Begin by drawing a 50 mm wide square in 123D, using the "Sketch rectangle" tool, following the grid on the work area.

With the "Offset" tool, create a second square, offset by 5 mm, inside the first.

Using the "Extrude" function, select the frame created and make it 50 mm high.

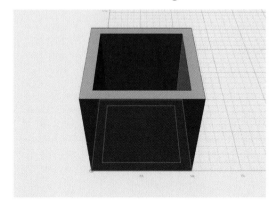

Use the "Sketch Rectangle" tool again to draw a 40 mm wide square centered on the front face of the hollow cube.

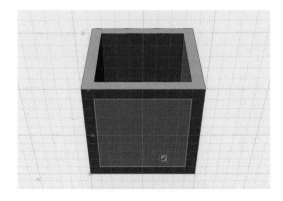

Select this new outline with the "Extrude" function, then by choosing a negative offset, extract the material throughout the object.

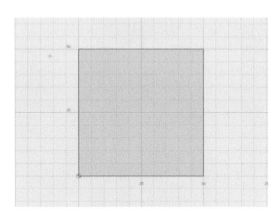

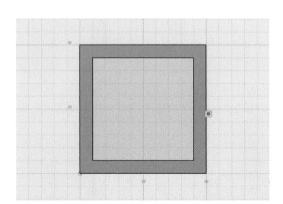

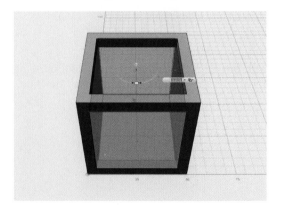

Repeat the last two steps for the side view. Only the edges of the cube will remain.

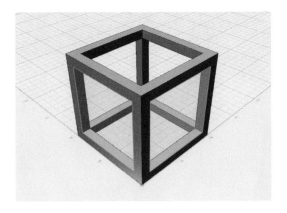

Add 2 mm rounded corners on each of the cube's inside edges. This sometimes help getting better bridges. Copy and paste the cube frame and scale

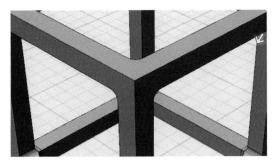

it down to 75%. Then copy and scale this new cube down another 75%. All three cubes will be inside of each other. Now, we need them to lay on

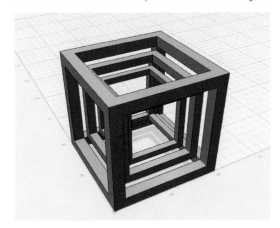

the same surface. For that, we will use the "Snap" tool and select the bottom surface of each cube.

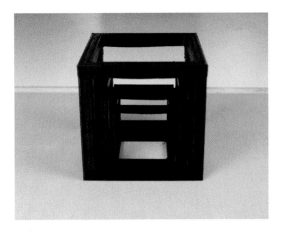

In your 3D printer's software, import your cube and copy it once with a scale of 75%. Then copy the new cube with an additional shrinkage of 75%, so that the three shapes can be nested. Make sure they are sitting flat on the work area. Some 3D printer software might not enable imbricated parts, when coming from seperate STLs. In this case, try doing it directly in 123D or Meshmixer. If you are a cheater, download the file directly from here:

http://www.thingiverse.com/thing:440375

Start the print and observe what happens at the end of each cube. Some filaments may float in the air, but normally each cube edge should be completed.

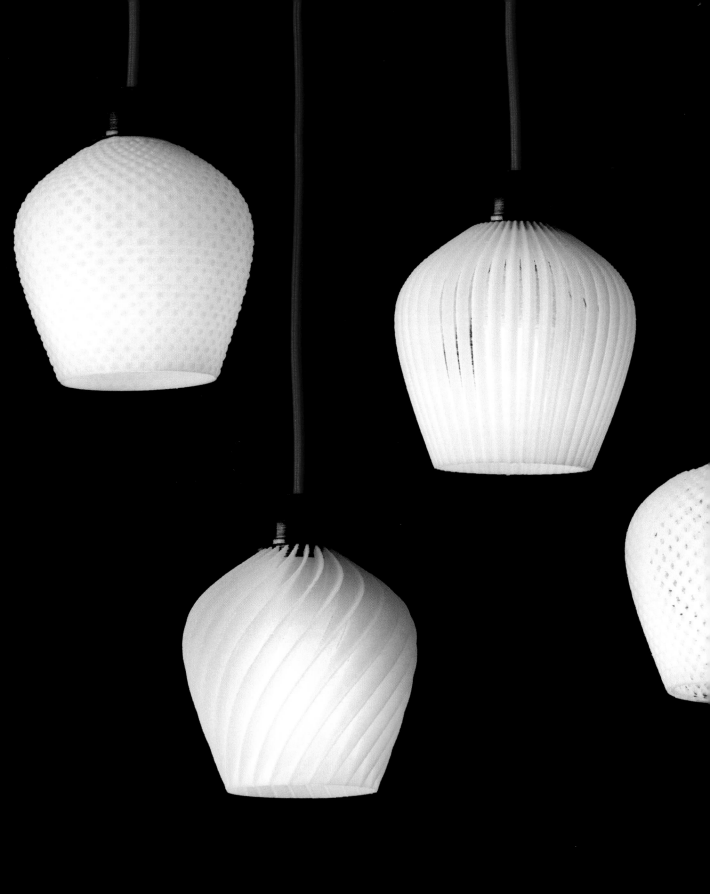

SET UP

AN EVOLVING DESIGN

PARAMETRIC APPLICATIONS

CREATE COMPLEX OBJECTS
USING SHAPESHIFTER : A LAMPSHADE

SHAPESHIFTER'S FUNCTIONS

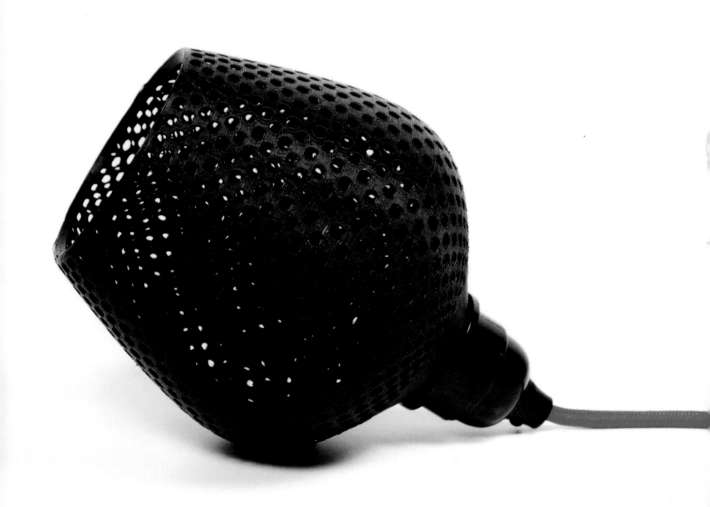

AN EVOLVING DESIGN

This collection of Dentelle lamps is proof that three-dimensional printing offers designers a hitherto unmatched flexibility of form. These little ABS shades, under 14 cm in diameter, were created in 2012 on the first generation of UP! PP3DP printers. The creator, an industrial designer who initially only wanted to repair an IKEA lamp in his living room, was caught in a creative whirlwind when he noticed the form possibilities his new tools offered. Modeled one by one using SolidWorks, a professional software package, the lamps quickly became a family with a lot of colors and textures. As soon as the designer's plastic experiment was shared on the internet, it was published on hundreds of blogs around the world, was a resounding success.

The idea of repairing and creating functional objects directly in your living room thanks to 3D printing began to take root with consumers. Today, with the help of 3D parametric software, there is no need to be a graphic or industrial designer to create shapes as complex as those in the Dentelle lamps collection. The designs can be infinitely modified and customized, simply by changing the data in some settings.

The exercise on the following pages will teach you how to use a parametric application to create your own shade using 3D printing.

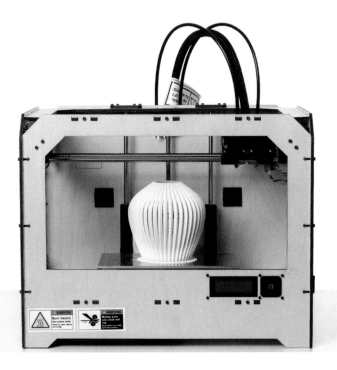

Previous page
Dentelle lamps
By: Samuel N. Bernier
3D printed on: UP! PP3DP
Source: Designer
© Véronique Huygues

Left page
Dentelle lamp
By: Samuel N. Bernier
3D printed on: Replicator 2
Source: Designer
© Véronique Huygues

Right page
MakerBot Replicator and Dentelle lamp
By: Samuel N. Bernier
Source: Designer
© Véronique Huygues

PARAMETRIC APPLICATIONS

Parametric applications are the future of bespoke design and "personalization of matter". Creating a parametric design application requires a good knowledge in computer science. To use it, however, is a breeze. We will see the kinds of applications that will be created in the coming years. Meanwhile, here is a selection of our favorites.

NERVOUS SYSTEM'S KINEMATICS BRACELET GENERATOR

The company Nervous System was created through the collaboration of an architect, passionate about biology, with a high-level computer expert. Together, they have developed several algorithms to automatically generate forms inspired by nature (fractals and others). Their research involved developing a jewelry collection, now famous in the 3D world. The duo (and couple) managed to put these algorithms in the form of a web application to give users the power to create their own Nervous System bracelets, necklaces, and rings. One of these interfaces lets users download STL file for free so that they can create and print in 3D themselves at home.

http://n-e-r-v-o-u-s.com/kinematicsHome/

MAKERBOT CUSTOMIZER

This free application allows any programmer, skilled in OpenSCAD, to create a modifiable design for users of the MakerBot Thingiverse community. It includes phone cases, paper embossers, and more. The tool is unfortunately limited by the use of OpenSCAD for creating editable files, but the application's use is straightforward.

http://www.thingiverse.com/apps/customizer

CHARMR

This application was launched by the company Autodesk on Valentine's day to customize a pendant by importing a JPEG or PNG image. The selected image is used as a "Bump-map" to generate 3D texture. When backlit, light passes through the jewelry's thinner places, giving the impression of a three dimensional monochrome image.

http://apps.123dapp.com/charmr/

SHAPESHIFTER

At first glance, Shapeshifter's website is not very impressive. On the screen, a simple purple prism is displayed on a sober blue background. It is only when you focus on the slide bars on the right of the screen that the magic begins. By changing the settings, the primitive shape gradually turns into a complex and refined object. This application seems perfect for generating vases, lamps, jewelry, and abstract sculptures quickly.

http://shapeshifter.io/

CREATE COMPLEX OBJECTS USING SHAPESHIFTER : A LAMPSHADE

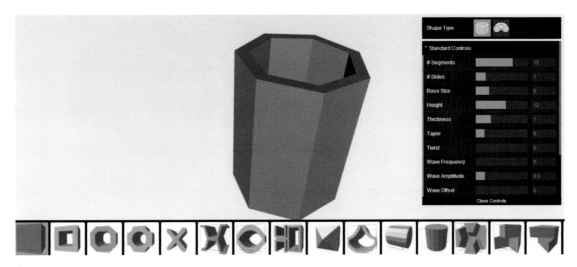

By understanding Shapeshifter's interface, you will be able to use the application to create complex objects easily. After you open their web page, you can see the following:

→ A menu at the top, enabling the model to be downloaded or shared.
→ In the center, a generic violet prism.
→ On the right, parameters to modify the generic prism.
→ At the bottom, different texture choices.

The application enables you to choose two basic shapes with their associated parameters: a prism or a Möbius strip.

From the menu, using the "Template" tool, you can choose an object template already set up to perform a function: vase, bowl, ring, bracelet, plate, chandelier, sculpture...

The "Randomize", tool unlike the "Template" tool, generates a completely random shape and randomly changes the settings sliders. The real interest of the application lies however in the control box and the generic volume setting.

Here we find the basic editing options which are accessible. They are detailed on page 92.

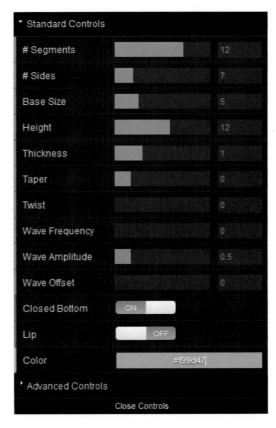

On Shapeshifter, it is therefore possible, from these simple settings, to create shapes as complex as those in the Dentelle lamps collection.

Here is how to do it. Open the application and its prism by loading the web page **http://shapeshifter.io/**

Enter the following parameters in the box of standard controls:

Segments : 16	**Taper : -0.89**
Sides : 24	**Twist : 0**
Base Size : 6.3	**Wave Frequency : 0.58**
Height : 9.43	**Wave Amplitude : 2.66**
Thickness : 0.22	**Wave Offset : 5.45**

You will get the general shape of the Lampshade.

Then uncheck the "Closed bottom" option and activate the "Lip" option. The lower surface should be open and the top hole reduced.

Closed Bottom : OFF
Lip : ON

Choose a texture from the bottom of the screen; forty that are available to you. Here we used the X-shape:

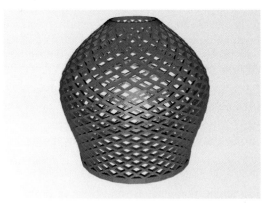

You should get a similar result to this one, but with whichever texture you chose.

If you wish, you can also give the shape a twist effect. We gave this one a 'twist' of 67:

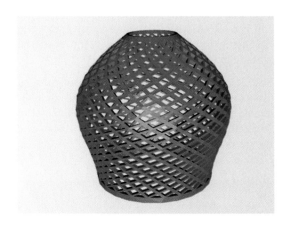

To export your creation, choose "Download Model" from the menu. Give it a name (with no accents or special characters), and then enter 132 mm for the Height. The other dimensions should be adjusted automatically. Keep the direction as Z and the OBJ file format, and then click Download. The file will automatically save to your download folder. If possible, check your model in Meshmixer before sending it for 3D printing. It should be possible to print the object without the need for rafts or support material.

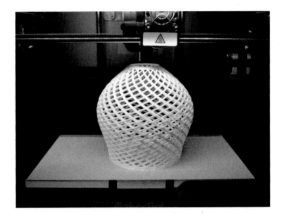

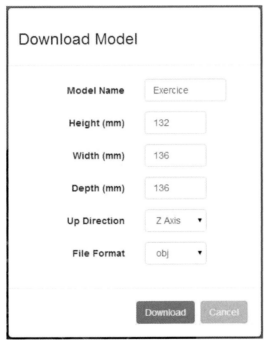

Download Model

Model Name	Exercice
Height (mm)	132
Width (mm)	136
Depth (mm)	136
Up Direction	Z Axis ▼
File Format	obj ▼

Download   Cancel

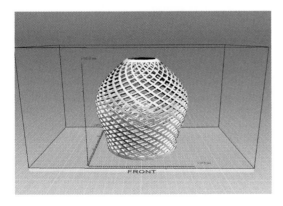

The top hole of the lamp is equivalents to a threaded socket for an E27 bulb. You'll need the standard plastic nut to secure it in place. For safety reasons, use only a low wattage LED bulbs.

Congratulations you have created a lamp using parametric software. Try now to replicate the same object on your usual 3D modeling software!!

SHAPESHIFTER'S FUNCTIONS

Sides : Changes the number of sides of the model.

Base Size : Changes the model's outer diameter.

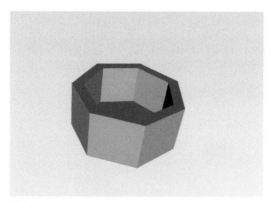

Height : Changes the model's height.

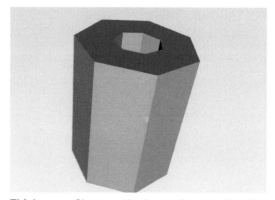

Thickness : Changes the inner diameter (i.e. the wall's thickness).

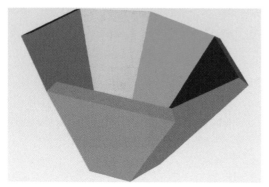

Taper : Changes the model's taper.

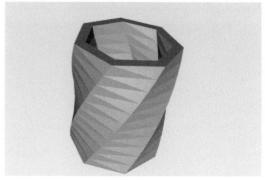

Twist : The volume twists on itself.

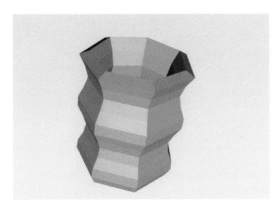

Wave Frequency : Creates waves on the surface, enables modification of the number of waves.

Textures : At the bottom of the Shapeshifter interface, you can select multiple textures generated by the application. There's something for everyone, from fish scales to delicately embossed netting. However, note that some geometries may be more difficult to print than others. Shapeshifter is still in the experimental software stage and it is not immune to mesh errors. It is therefore important to check the status of your files after exporting a repair software such as Meshmixer or Netfabb.

Wave Amplitude : Changes the amplitude of the waves.

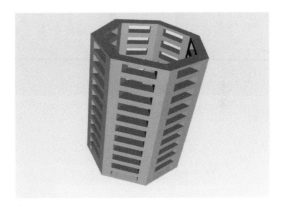

Square texture.

Wave Offset : Offsets the waves over the height of the model.

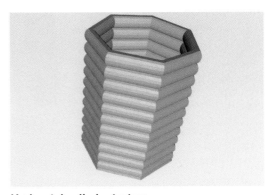

Horizontal cylinder texture.

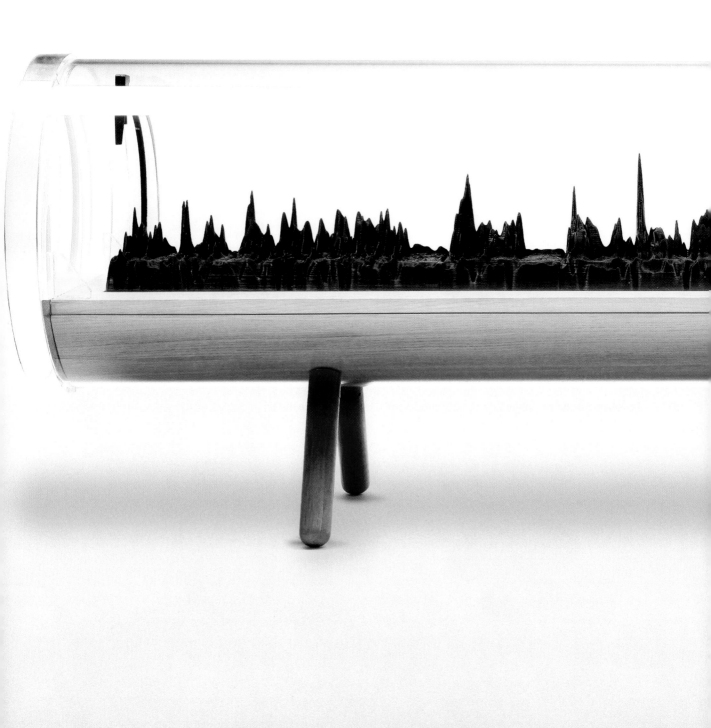

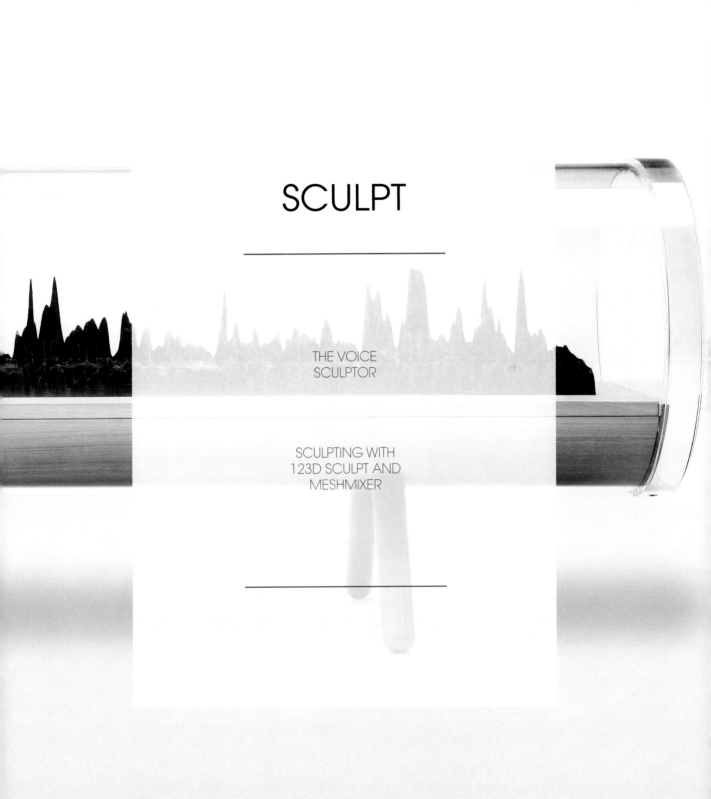

SCULPT

THE VOICE
SCULPTOR

SCULPTING WITH
123D SCULPT AND
MESHMIXER

THE VOICE SCULPTOR

Industrial designers, engineers, and architects do not have a monopoly on 3D printing. The work "Barack Obama Next Industrial Revolution" is proof. Produced by Gilles Azzaro, a digital artist from Casablanca, this interactive sculpture materializes the invisible. Azzaro is fascinated by sound and this is not his first work using voice as the medium. However, with this creation of the US President's voice-print from the February 2013 "State of the Union Address," where Obama claims that 3D printing will head the next industrial revolution, the artist has entered the annals. In fact, the 3D printed sculpture was received by Obama himself during the White House Maker Faire in 2014.

The aesthetic of Gilles Azzaro's work is unique in itself. The artist has deliberately used a very low print resolution on his sculpture to accentuate the strata effect by creating thick plastic layers.

To generate a three-dimensional interpretation of the voice, Gilles Azzaro uses various personal and patented technology. The peaks are created by algorithms related to the voice's pitch.Several modeling software packages enable you to actually sculpt virtual material. ZBrush, Autodesk Mudbox, Cinema 4D, and Sculptris are the most used among professionals. Now there are also free solutions for PCs and iPads which enable intuitive creation of shapes. Meshmixer can merge, smooth and distort shapes from different sources while 123D Sculpt, 123D Creature (now merged in 123D Sculpt+) use touchpad tablets to take maximum advantage of natural hand movements. The following exercises show you how to use them.

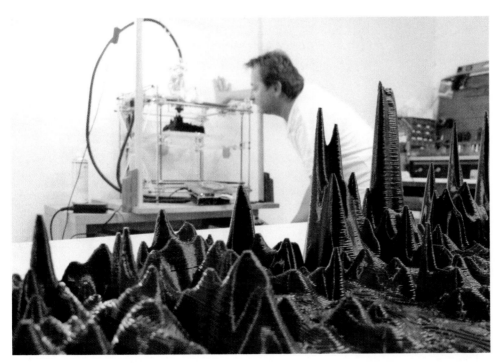

Previous page
Sculpture: Barack OBAMA:
The Next Industrial
Revolution
By: Gilles Azzaro
Source: www.gillesazzaro.com/
© Gilles Azzaro

Left page
Barack OBAMA: The Next
Industrial Revolution
By: Gilles Azzaro
Source: www.gillesazzaro.com/
© Gilles Azzaro

Right page
Gilles Azzaro in his lab
By: Gilles Azzaro
Source: www.gillesazzaro.com/
© Gilles Azzaro

SCULPTING WITH 123D SCULPT AND MESHMIXER

SCULPTING A MODEL WITH 123D SCULPT

For this exercise, you'll need an iPad tablet, equipped with the free application 123D Sculpt. You must also have created a free account on the Autodesk 123D community.

We will try to recreate "manually" the stalagmite effect present in Gille Azzaro's voice sculptures.

To begin with, open the 123D Sculpt app on your iPad. Create a new project by pressing the "+" symbol. This takes you to the shape templates library. There you will find the following families: Creatures, objects, and geometry. First, there are animal and human body forms. Second, primitive forms and third, vehicles, clothes, and other miscellaneous objects. Select "geometry" and choose the cube by touching it with your finger.

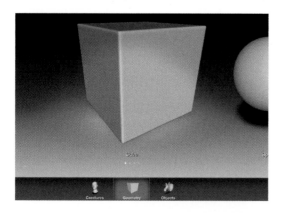

Place your cube in a forward facing view so that you see the top surface with a minimal incline.

Using the material removal tool (in the left column, eighth from the top), represented by a sphere and an arrow pointing outward, pull the surface above the cube. Decrease the tool's "Size" to 30, then adjust the "Strength" to 100%. You should then have created a protrusion that looks like a stalagmite.

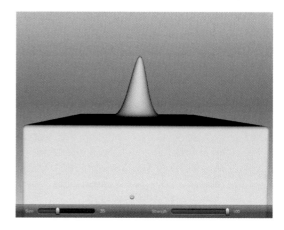

Vary the size and the strength of the tool to get a random look. Repeat several times over the entire surface, but leave some empty areas for the next tutorial.

When you are satisfied with the shape use the other sculpting tools to soften, flatten, or thicken the tips.

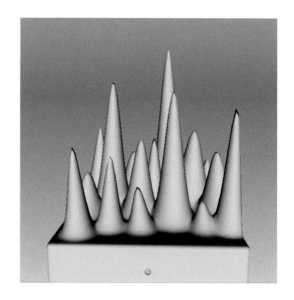

Save your model in your sculpture gallery, and send it to the 123D web community using the "Share" button. You will have to give your work a title.

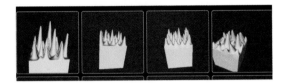

Your model can be in either "Public" or "Private" mode.

From your computer, go to the **www.123dapp.com** site and log on to your account. Under your "Me" avatar at the top right, click on "Models". Your latest creations should be displayed. If they are not there, make sure that your tablet is connected to the Internet or wait a few more minutes. Select the shape created on 123D Sculpt, then go to "Edit / Download". There are several options, including "Order a 3D print", "Send to 123D Make," and "Print at home."

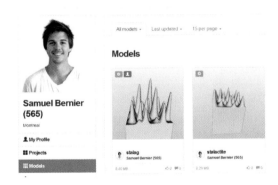

Click instead on "Download 3D models". A document in ZIP format containing an STL file and a colorful OBJ file should load onto your computer.

If the download options are displayed in gray, wait a few minutes. Your model might not have finished loading online.

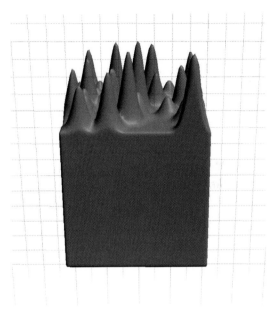

SCULPTING A MODEL ON MESHMIXER

Open the STL file you have created in 123D Sculpt in Meshmixer. If you couldn't make the previous exercice, you can always download the example file named "Sculpt exercise" from:

www.thingiverse.com/thing:442538

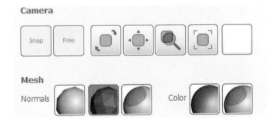

When your file opens, you can change the display options by pressing the "Space" key on your keyboard. You can choose your interface's background color, and the "Mesh" display type.

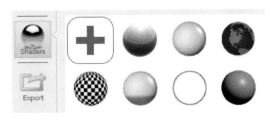

You can also choose display colors and textures for your object using the "Shaders" tool.

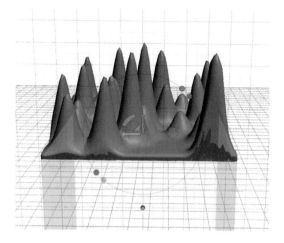

With the "Plane Cut" function in the "Edit" tool, remove the remaining mass from the cube under the stalagmites.

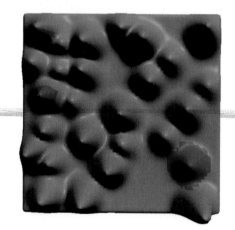

Next click the "Select" tool and then, using the brush, select an area covering one of the peaks. The left window will change to show the new options. Go to "Convert to" and then "Convert to Open Part". The shape will automatically be saved to your "Meshmix" library.

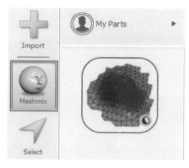

In Meshmix, go to "My parts", then select the shape you've just saved. Slide the shape so that it fills the holes you left in the previous exercise.

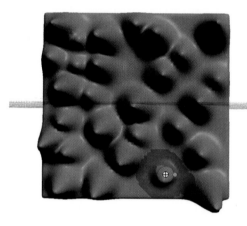

The object as shown on the screen, is only 1 mm wide in reality. In the "Analysis" tool, select "Units / Scale" then change size X to 60 mm. The rest will adjust automatically.

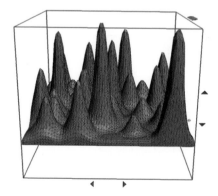

Now, in the "Sculpt" tool, use the "Brush" called "Refine" to increase the density of the mesh areas where the polygons are most visible. To help you visualize your creation's mesh go to the "View" menu, then "Toggle Wireframe" (keyboard shortcut W).

Open it using your 3D printer's software and start printing at a very low resolution (0.3 mm to 0.5 mm) to accentuate the layered effect. Supports and rafts are not needed.

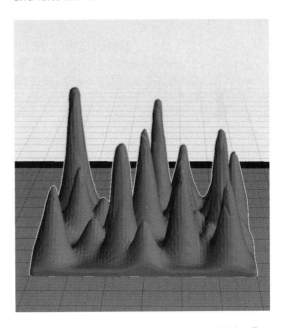

Here is the result of a 3D print on a MakerBot Replicator 2 with 0.4mm layers.

Note that from the "Sculpt" menu, you have access to almost all the modeling tools in 123D Sculpt. The "Drag," "Spike," and "Move" tools, for example, make it perfectly possible to create stalagmite forms similar to the ones we made on 123D Sculpt. It is therefore possible to use Meshmixer for sculpture, but without the tactile advantage of a touch screen tablet.

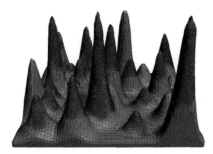

Run "Inspector" on your model to look for any repairs required before exporting your new STL file.

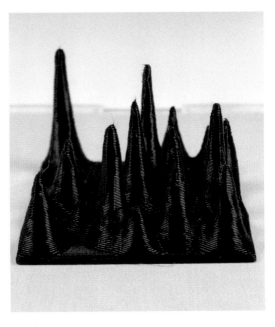

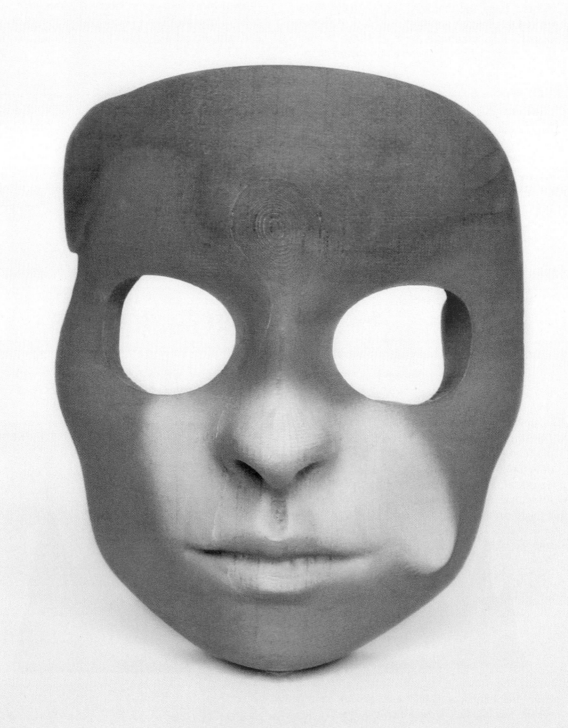

COLOR

_____

A NEW APPROACH TO COLOR

COLOR WITH 123D SCULPT
AND MESHMIXER

TIP:
TO GET MULTICOLORED PARTS

_____

A NEW APPROACH TO COLOR

Laureline Galliot is one of the few designers to have experimented with modeling and virtual coloring whilst creating everyday objects. With her Contour & Masse project, she presents a modeling method which uses aggregated solids on which she applies colors and textures from her personal palette rather than exploiting lines and extruded surfaces, as is the norm in product design. With her extensive painting experience, Laureline Galliot has used digital sculpting tools intended for animation, to paint, model, and intuitively modify her objects.Combined with color 3D printing techniques such as "Binder jetting" 3D Systems (Z Corp) and SDL paper from Mcor IRIS, Laureline Galliot's creative processes have brought to life objects as colorful as they are original. This new approach to object creation gives new freedom to the use of color in product design.

Spotted the Villa Noailles in 2013, her project won the prestigious prize from the Var General Council.

Inspired by Laureline Galliot's creative process, the following exercises demonstrate how to use free software for coloring and modeling shapes for 3D printing.

Again, there is a multitude of professional software packages that enable you to color, texture, and "map" visible file surfaces and to send them for 3D multicolor printing.

Here, we decided to use the same two free tools that helped us to sculpt in the previous exercise. This tutorial will teach you how to use the color functions in 123D Sculpt and Meshmixer.

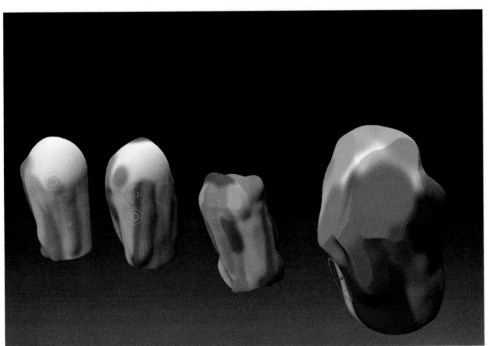

Previous page
Contour de masse, Mask
By: Laureline Galliot
3D printed on: Mcor Iris
Source: Laureline Galliot
© Laureline Galliot

Left page
Contour de masse, Thea pot
By: Laureline Galliot
3D printed on: ZPrinter 650
Source: Laureline Galliot
© Laureline Galliot

Right page
Designing the tea pot
By: Laureline Galliot
3D modeled on: ZBrush
Source: Laureline Galliot
© Laureline Galliot

COLOR WITH 123D SCULPT AND MESHMIXER

COLOR WITH 123D SCULPT (IPAD)

Import a generic sphere into the application from the 'Geometry' library. Using the digging tool, third from the top, drill small holes on the surface of the sphere, as if you were trying to make a golf ball. Reduce the tool's size and strength to 50%. To save time, you can use the symmetry function so that your movements are mimicked on the opposite side of the sphere.

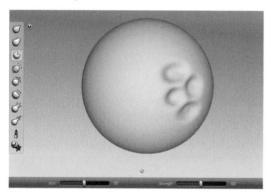

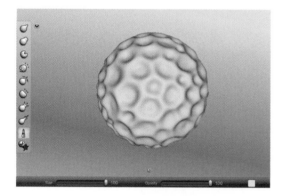

When your ball is finished, select the "Brush" tool. The second to last icon in the left hand column.

Choose a background color for your sphere. Use the circle as your brush shape, then allocate its thickness and maximum size before applying color over the whole surface. Still using the "Paintbrush" tool, change the tip to the blurred circle, choose a complementary or contrasting color, and use that for the base of the sculpture. Reduce the size of the brush to 20%, then fill each cavity with color, alternating the color tone now and then.

In 123D Sculpt, you can also apply images and textures on an object's surface using the "Mapping" tool (at the end of the right hand column). Visuals like eyes, faces, skin and materials are there by default, but you can also import your own images from the iPad's camera or photo album.

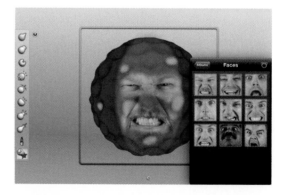

To apply the selected texture, frame the picture on your object and then slide your finger on the screen to where you want to apply it. You can also adjust the opacity. When you are satisfied, save and then share your sculpture on your 123D account from the gallery, and then retrieve it on your PC. The ZIP file will contain your OBJ model along with the MTL file and the associated PNG mapping.

All that remains is to print it on a color 3D printer!

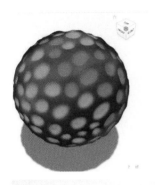

COLOR WITH MESHMIXER

Import a sphere in Meshmixer using "File, Import Sphere".

In the "Sculpt" tool, use the "Brush" called "Drag" pushing it onto the sphere's surface to reproduce the crevasse effect that we made using 123D Sculpt. Set the "Strength" to 30 and the "Size" to 50. Make use of this step to explore the effects of the different brushes on the sphere.

Before moving on to the color, make sure the "Mesh Color Mode" is enabled by pressing the space key and selecting the sphere painted blue.

In the "Sculpt" tool, select the "Paint Vertex" brush.

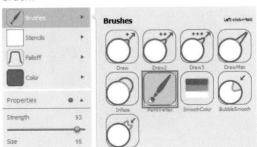

For your "Stencil", select the last one on the right, depicted by an indefinite circle. Note that you can create your own shape by importing an image using the + icon.

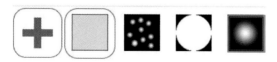

In "Falloff" select "Bubble", the fourth icon.

Feel free to switch from one mode to an other to discover the effect of each tool.

Paint the entire surface of your sphere, choosing a color and a large brush size to create a background. The color of the shader will not affect the hue of the file once exported.

Change the color and decrease the size of your tool so that it matches the diameter of the cavities you produced.

By changing your brush to "Smooth Color", in the surface mode, you can attenuate the transition between colors and even combine them. You can choose the size of your model before you export your sculpture to avoid unpleasant surprises at the

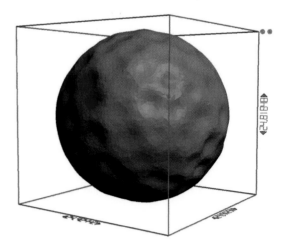

time of production. Go to "Analysis » Units" and enter the dimensions that you want. By default, the sphere is 25 mm in diameter. To save a color file of your sculpture on your computer, go to File » Export » OBJ Format With Per-Vertex Color. Unlike 123D Sculpt, the OBJ file will not be associated with an external MTL file, because all the color informations are already embeded in the Per-Vertex OBJ. If you want to have a separate texture file, export with a standard OBJ format.

TIP:
TO GET MULTICOLORED PARTS

CHANGE THE FILAMENT (PAUSE Z)

It is possible to change the filament during your print, either by monitoring your part and pausing the process to reload, or by configuring the machine, if a z pause option is available. Some slicing software, like Cura and Simplify 3D, also offer the possibility to set multiple Z pauses directly in the 3D printing code. You can change the filament color as many times as you want and create multicolor pieces with a single extruder. While you're at it, try also to change the type of material while extruding. Go from PLA to NinjaFlex to wood filament and see what happen. You might discover a new composite material.

NYLON AND DYE

Nylon filament can be bought in certain stores: its advantage, outside strength and flexibility, is that it can be colored with textile dyes. This kind of dye comes as powder and needs to be mixed in hot water. Follow the instructions provided with it, as if you were working with textile. Make sure your dye is nylon compatible. Immersing your object in a container filled with diluted dye will color it in about an hour. You can also dye the filament directly using different colors on the same spool to create hue transition on your 3D printed objects...

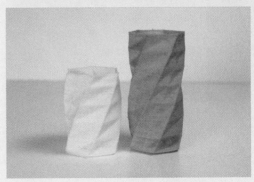

3D SCAN

SOUVENIR OR MEGALOMANIA?

3D scanners have been used in industry for decades for reverse engineering, that is to study the geometry and wear of existing objects. They are used in architecture and urban planning to improve cities and to preserve our cultural heritage. Scientists themselves use this technology to create a virtual archive of valuable objects and to make reproductions to be used when studying certain artifacts. In medicine, 3D scans enable lives to be saved and increase quality by creating custom prostheses and even implants.

What will consumers use them for? For now, there is a one word answer: "selfies". These self-portraits (or "egoportraits") have been invading modern-day social networks since 2004 and now they come in 3D. With mobile photogrammetry applications, such as 123D Catch, you can make a three-dimensional scan anywhere, anytime. The 3D selfie will become even easier when infrared LED scan techniques are directly integrated with our handheld devices. Google is already on the ball with its Tango project, a telephone that creates point clouds of spaces in real time. In 2013 Apple bought the company PrimeSense, the originator of the technology in the first generation Kinect for Xboxes and holder of multiple patents on the use of infrared LEDs.

3D photo booths are also fashionable around the world. The Japanese company PARTYLab was one of the first to make this concept popular through the Omote 3D Sashin Kan project. This publicity stunt, sponsored by the 3D iJet printers, had an immediate response from international media. The idea was used by Coca-Cola in Israel and Asda, the English branch of Walmart, for promotional purposes. The figure on the left hand page was created in France by the 3D photographer Dominique Lentengre using a 3D Artec EVA scanner.

Previous page
Place de la République
Scan: Trimble
By: Grégory Lepère
Source: le FabShop
© le FabShop

Left page
Mini-me
Scan: Artec EVA
By: Comexx
3D printed on: ZPrinter 650
Source: Comexx
© le FabShop

Right page
Pink and blue miniatures
Scan: Asus Xtion with Skanect
By: Samuel N. Bernier
Source: le FabShop
© Samuel N. Bernier

SCANNING AN OBJECT

WHY 3D SCANNING?

There are 3D scanners in all price ranges with a corresponding quality range; in general, the more you spend, the more faithful the reproduction. However, no single scanner has the solution to every problem. Before you invest time and money in scanning, you need to be able to determine whether an object should be digitized with 3D scanning or manually modeled with the help of precise measurements. Here are some cases where the use of a 3D scanner is justified.

HANDMADE ITEMS

Some shapes are just a headache to model. They might be too organic to be measured manually, or have a surface texture that is too random. Handmade objects are ideal subjects for 3D scanning. Are you a sculptor? Digitize your creations to create a virtual portfolio. That way you will have a digital record of your work.

SCALE

It is possible to digitize virtually any dimension of object from entire buildings to microscopic ones (with the appropriate tools). A 3D file can change scale infinitely, which lets you reproduce the digital subject in any desired size. Changing an object's scale has never been so easy, but be careful, you may lose details once the file you produced is reduced and 3D printed.

CUSTOMIZATION

3D scanning is used in medicine and in fashion to take the exact measurements of a subject in order to create a custom accessory. For example, prostheses for people with one leg can be made from a scan of the intact opposite leg. The US army is also studying the use of 3D scanning to design armor adapted to the soldiers anatomical differences.

PRECISION

In industry, high-accuracy 3D scanners allow verification of the wear and deformation on the manufactured components. The 3D scanner is therefore more of a tool for testing quality than a modeling tool.

REALISM

3D scanning preserves a level of realism that is difficult to obtain with 3D modeling. Even the movement of fabric, the posture of a person, or the expression on a face. Moreover, big movies with animation and special effects, such as Avatar or The Lord of the Rings, use techniques similar to 3D scanning to give digital characters more realism.

FRAGILE OR PRECIOUS ITEMS

3D scanners are used in archeology and conservation to produce replicas of priceless artifacts. This can be used to create hands-on exhibits without risk of damage to the original, or to share with schools around the world to produce their own copies for use in the classroom.

THE PREREQUISITES FOR THE OBJECT TO BE DIGITIZED

OPACITY

In order to be able to digitize an object in 3D, the object must be able to reflect light. Some transparent materials, such as glass, simply cannot be digitized in their natural state. In order to be able to calculate the sides of an object with a certain translucence, we use a "mattifying" agent, often in the form of powder or a liquid in the form of an aerosol. The treatment consists of covering the object with a fine opaque layer that you can remove with water or compressed air. A more drastic strategy consists of completely repainting the object.

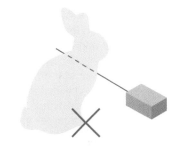

MATTE

The same problem with reading digitized surfaces also applies to objects that are too shiny or reflective. A mirrored surface, polished metal or crystal can't be digitized because of its shiny surface. The reflections and refractions of of the light confuse the reading of the geometry and prevent the calculation of a coherent volume. In order to counteract this problem, using a mattifying agent is also suggested. Of course, bright objects, such as a light bulb or a white surface exposed to direct sunlight pose the same problem of scrambling the data.

THICKNESS

Digitizing a flower is not impossible, but it's still somewhat complex. When the volume to be scanned is extremely thin, the scanner can have difficulty determining the front side from the back side. For this reason, it risks canceling the geometry or superimposing the polygons. This is particularly true for medium and low-resolution scanners such as the Sense from 3D Systems. The problem with thickness also applies to opacity.

STABLE AND STATIC

The object and its environment have to be stable for the entire duration of the scan. Likewise, an exterior element that goes into the digitization area interferes with the meshing of the 3D file. The majority of 3D scanners calculate the geometry on the inside of a virtual box. For the MakerBot Digitizer, this volume is a cylinder that is 20 cm in diameter by 20 cm high. For the Xtion scanner, this virtual area is an adjustable cube that can go from 30 cm to 500 cm per side. All geometry collected outside this border will be automatically ignored.

LIGHT-COLORED ITEMS

With the use of infrared LED scanners, very dark-colored objects are problematic since they absorb the light emitted rather than reflect it. In this case, the use of a mattifying agent or talc is a solution. Dry shampoo is also frequently used.

TEXTURED ITEMS

In the case of a color or photogrammetric scan, the textures can serve as a point of reference, so they are useful. With applications such as 123D Catch, it is also suggested that you add markers before you scan something with different symbols to make it easier to calculate data collected from volumes that are too even in shape or in color. These "markers" are often made from numbered Post-it notes or pieces of painter's tape.

ORGANIC

A digitized cube will never be more precise than it is in a 3D modeling software, and this also applies to a screw, a door hinge, or drawer slides. An object whose surfaces are measurable and even can, of course, be digitized with the help of a 3D scanner, however, if it can be modeled simply, modeling remains the best solution. Volumes with complex curves are ideal subjects since they are more difficult to measure by hand.

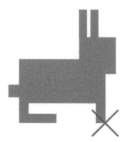

PREPARING THE SUBJECT FOR 3D SCANNING

Begin by covering any transparent or shiny object (glasses, silver jewelry, piercings…). Use dry shampoo, available in any pharmacy, to cover surfaces that are too shiny or too dark. Stabilize your sub-

ject. If the subject is a person whose bust you would like to scan, ask him or her to sit down. If

it's an object that is unstable, use adhesive putty or adhesive tape to hold it in place. Some people use a rotary table so they can pivot the subject while keeping the scanning tool in place. If you are trying to digitize a live animal… well, let's hope that it's well-trained.

TIP: DIGITIZING MULTIPLE TIMES

MULTISCANNING

The range of 3D scanners is limited. You sometimes have to resort to tricks to reproduce the largest and most complex objects. Some 3D scanning software programs, such as MakerWare for Digitizer, directly integrate a "Multiscan" option, which allows you to digitize the same item several times by swiveling on different axes between each scan.

COMBINING FILES

If you use solutions such as Skanect, you will have to bypass the problem by saving several "paths" that you then recombine with the help of the "Reconstruct" tool. You can also simply save multiple STL files that you then merge manually using Meshmixer.

USING SKANECT
TO CREATE A MINIATURE BUST

For this exercise, you should get your hands on a Microsoft Kinect 1 sensor or an Xtion Pro by Asus, two accessories that use PrimeSense technology. If you have a Sense, iSense, or Structure sensor, the procedure will be similar, but the software interface will be different.

Download and install the free version of Skanect, or buy a pro license for $129 from:

http://skanect.occipital.com

Plug your device into the USB 2 port of your computer. Make sure that all of the drivers have been installed before beginning.

After having prepared your subject, create a perimeter of 1.5 m around it and make sure that there is no direct light source that can scramble the 3D scanner's sensors. With the Skanect software, go into "Prepare » New". Choose the standard stage called "Object." By default, it uses a virtual box that is 0.6 m on each side, which is perfect for our subject. Keep "Aspect Ratio" in normal mode. The "Height x 2" mode corresponds to a vertical box, very useful for digitizing people from head to toe.

Press "Start" to begin.

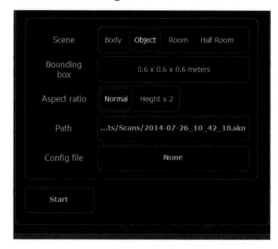

Orient your infrared scanner towards the subject. It should appear in the box that is displayed on the screen. Adjust your distance in such a way that the head and shoulders of the person you are digitizing are entirely in this box. Help yourself by pivoting the cube on the screen.

When everything is in place, click on the black circle on the red background at the top left. A countdown will begin. This can be set with the help of the cursor to the left of the black circle.

Your subject should be displayed in green and red on the screen. The green surfaces correspond to the geometry calculated, the red ones to areas out of range. Carefully pivot around the subject or rotate the item itself. Don't forget to scan the top of the head and under the chin.

If you lose your place along the way, try to super-impose the gray image with what is displayed at the bottom of the screen. If that doesn't work, cancel the operation and click on the circle to begin again from the start.

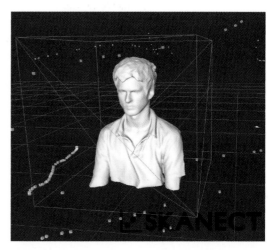

When you have finished, click on the red square at the top right. After several seconds of calculating, the 3D volume will be shown in white.

Go directly to the "Process" tab. There you will find automatic repair tools, but we won't be using them. Meshmixer lets you repair everything just as well, but with more control over the results. You can still position and cut the undesirable areas of the file with the help of the "Move and Crop" tool.

Make sure that your subject's chin is pointed slightly up, which will allow it to be made without support material.

If your Kinect or Xtion is equipped with a webcam, you can retrieve the colors by clicking on the "Colorize" option.

In the "Share" tab, choose "Export Model." You will have the choice between several formats and units of measurement In format, choose OBJ. In "Colors," choose UV Texture to get an MTL file. In number of faces, export with the maximum unless you want to create a faceted effect. If you are using the free version of Skanect, you will only be able to export a limited number of surfaces. Use millimeters as the unit of measurement in "Scale" mode. That way you will get a color 3D file at original scale that can be used with multiple 3D printing machines.

Now open the file in Meshmixer and launch an automatic repair procedure with the "Inspector" assistant (See the exercise on page 43).

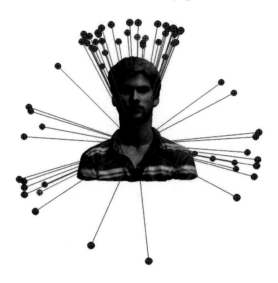

Make sure that the base of the object is very flat and closed. If not, use the "Plane Cut" tool. Your bust is ready to be printed in 3D, but first add a base to it.

In 123D Design, create a base by using the sketching tools and the "Revolve" function, then export an STL file that you imported to the scan open in Meshmixer by choosing "Append." Don't worry about the size of the base, we'll adjust it in Meshmixer.

Import the STL file into the slicing software of your 3D printer. The object should appear at its original scale, therefore much bigger than your print volume. Give it a scale of 1:10, which should reduce it to 60mm in height.

In the "Edit" tool, use the "Transform" function to adjust the size, depth and position of the base. Make sure that it makes contact with the bust. In the "Object Browser" window, select the two volumes to make the boolean options appear. Choose "Boolean Union" to fuse the volumes. Export the result in STL Binary format. If the sculpture is not personal enough for you, return to 123D Design and add a name to the base by using the "Text" tool. Then begin the previous procedure again.

Start the print with maximum resolution. If you would like, you can create a mold of the model afterwards for the color in plaster or metal. The same file could also be used for digital milling in high quality materials such marble or wood, if you are into this kind of decoration.

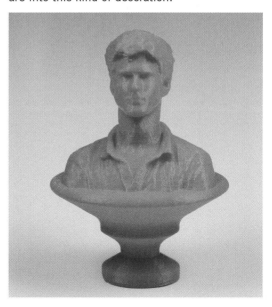

USING 123D CATCH
TO CREATE A 3D SCAN

Begin by installing the 123D Catch application on your Apple or Android mobile device. A PC compatible version also exists, but it requires the use of a digital camera and behaves different than the mobile version described here. You have to log into your 123D account to use all of the application's features.

www.123dapp.com/catch

Choose a textured subject that is non-transparent and stable. We have decided to digitize a stuffed toy tiger. Its stripes are ideal for serving as a reference for the photogrammetric process.

The texture of the surface on which the object is placed can also help to obtain a precise model. Here, the wooden slats and the Persian rug below have contributed to the success of the scan. Take photos of the object from all possible angles. A mark on the lower left helps you determine the necessary number of pictures. 25 photos should be enough for a usable model. When you have finished taking photos, touch the check-mark on the top left of the application. Validate the rest of the photos by pressing "Submit" at the top right. The calculation begins, and could take several minutes or several hours. It depends on the speed of your Internet connection and the traffic on the server. Memento, an even more powerful photogrammetry tool by Autodesk, allows you to get an even better results by analyzing more high resolution pictures at the same time, which can take more than a day. See **memento.autodesk.com**

You will receive a message when the 3D reconstruction is finished. You can then visualize your virtual object and verify if the digitization has gone

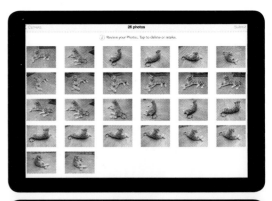

well. In the case of the tiger, only a hole under the chin poses a problem, but we can repair it later with Meshmixer.

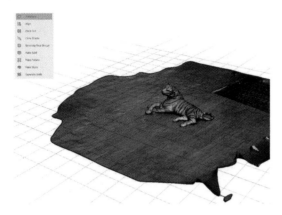

Connect to your 123D account on your computer to recover the 3D file from your "Models" library then open it in Meshmixer. Using the "Select" tool, select all of the undesirable space surrounding your subject. The area should appear in orange. Press the X key on your keyboard to make it disappear. Then,

go into the "Analysis" tool and pass the volume to the automatic inspector to seal all of the holes. Make sure that the bottom of your object is very flat using the "Plane Cut function", then export a file in STL Binary format. Export an OBJ instead if you want to produce it in multiple colors on an

Mcor Iris or a Projet 660 by 3D Systems. Find the result of this tiger 3D scan here: **www.thingiverse. com/thing:440462**

Import the cleaned file to your 3D printer's software, then adjust it to the desired scale. The level of detail obtained is not enough to create a large-format

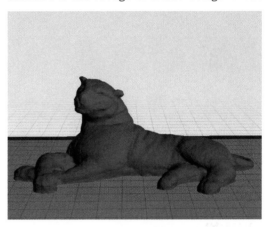

object, unless you sufficiently rework the file with software programs such as ZBrush, or Mudbox. We printed our little tiger without raft or support and with an infill density of 10%. Photogrammetric scanning is a great tool, especially when it is combined with our mobile devices. It lets you generate digital copies of both large objects and small ones with unequaled color precision. However, a good dose of patience and training is necessary to obtain high-quality results, but the answer often lies in the choice of subject and environment.

OPTIMIZE

THE SIZE OF FILES

LOCALLY REDUCE AND ENLARGE
THE MESH WITH MESHMIXER

REDUCING THE SIZE OF A FILE
WITH MESHMIXER

THE SIZE OF FILES

The type of faceted/polygonal surface that is found on some pieces made with 3D printing is mostly due to a 3D file with too low resolution.

In the world of 3D, the polygon is to the STL file what the pixel is to the JPEG file. The number of polygons in a file will determine if part of its surfaces will be smooth or faceted. For example, a sphere without enough polygons will look like a disco ball. This principle doesn't apply to flat surfaces since they can be defined with a tiny number of polygons. The number of facets in a 3D file can be managed directly with modeling software, at the time of export or in the display settings.

In the case of the Muse corset, a scan of the model was created to serve as a basis for the 3D modeling. A file resulting from a high-resolution 3D digitization can easily take up a gigabyte of memory space. In order to be imported into modeling software without a problem, the 3D scan needs to be simplified as much as possible to keep only what is essential. We call this *mesh decimation*.

Since the Muse corset has very organic curves, an impressive number of polygons is necessary in order to reproduce the model's shape. Not only was the 3D file generated by the modeling software excessively heavy, but the capacity of the laptop being used was pushed to its limits.

Although it's possible to regulate the density of the mesh of an STL file when you export it in several software modeling programs, it isn't the case for all of them. The following exercise will teach you how to reduce the size of files that are too large and choose the best mesh ratio based on the scale of the object.

Previous page
Corset muse
By: Samuel N. Bernier
3D printed on: Replicator 2
Source: Thingiverse
Thing: 245461
© Samuel N. Bernier

Left page
Corset muse
By: Samuel N. Bernier
3D printed on: Replicator 2
Source: le FabShop
© Samuel N. Bernier

Right page
Scan of the subject
By: Samuel N. Bernier
Source: le FabShop
© le FabShop

LOCALLY REDUCE AND ENLARGE THE MESH WITH MESHMIXER

Import a sphere into Meshmixer. In the "Sculpt" tool, choose the "Reduce" brush.

Adjust the strength of the tool to 90 and do the same with its size.

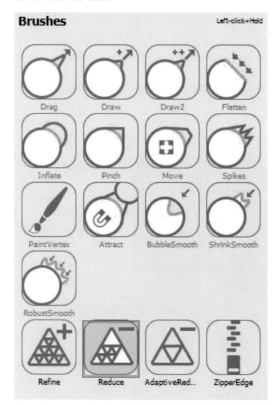

With the help of your mouse, facet half of the sphere in such a way that the triangles are very obvious. To help, enable the display of the mesh (Wireframe) with the W key on your keyboard.

For the second half of the sphere, use "Refine." This tool will increase the mesh's density to the maximum.

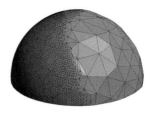

Cut the sphere in half with the "Plane Cut" function in a way that keeps the two resolutions visible. Export the file in STL then import it into your 3D printer's software. Copy it three times. At a scale of 100%, 150%, and 200%.

Print the half-spheres with infill of 0% in full resolution (0.1 mm for most FFF 3D printers). You will observe that, although the change in the mesh is very visible on the half-sphere size at 200%, is is barely perceptible on the smaller piece.

You can understand the importance of calibrating and choosing the resolution of the mesh depending on the size, the geometry resolution, of the layers that the technology being used allows. If you are making the same three volumes with a DLP 3D printer with 16 microns of resolution, the bumps on the surface will be visible on each of the pieces.

REDUCING THE SIZE OF A FILE WITH MESHMIXER

For this exercise, you will be downloading a very high-resolution 3D model. We have chosen a hand digitized with an Artec EVA scanner. You can recover the 243 MB STL ASCII file, named "main_HD_cut_ascii.stl", at the following address:

http://www.thingiverse.com/thing:440625

Open the file in Meshmixer. Before doing any modification, export it again, but this time as an STL Binary format. The size of the file should automatically be reduced to 44 MB, which is 6 times smaller than the original file.

Now, go into the "Select" tool and press the Control-A (or Command-A) keys to select the entire volume. Its color should automatically change to orange. Activate the "Wireframe" to help you visualize the grid (W shortcut).

In "Edit," click "Reduce" or use the shortcut (Shift-R). First reduce by 50%, then export the result in STL Binary. The size of the file obtained should have been reduced to 22 MB.

Repeat the experiment with a resolution that is about 50% lower. Your file should now be around 11 MB. Study the transformations hapenning to the hand's mesh. Make an additional reduction of another 50% to the 11 MB file. The size of your file must be less than 6 MB. Your original file has been divided by about 40. A fourth reduction by 50% will give you a file with the very acceptable size of about 2.5 MB.

The difference from the original file is still very subtle. With a fifth and final reduction, this time by 75%, you will begin to see a distinctly more faceted texture appear.

Your file is no more than 0.6 MB, 400 times smaller than the high-resolution STL ASCII. You can now visually compare your 243 MB file to the most recent obtained by importing them into the same project.

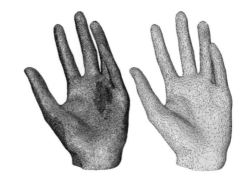

Do you see a difference?

If yes, import the two files into your 3D printer's software. Begin a 3D print by using the same machine, the same consumable and the same resolution for the two pieces. You will notice that they have almost exactly the same appearance once you've printed them.

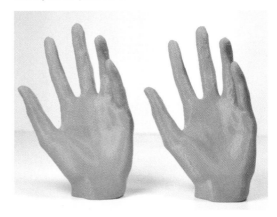

We rarely need large files to create objects with FDM/FFF 3D printing. However, it is interesting to have tighter mesh when you want to use more precise technologies, such as DLP or when you are creating photo-realistic images. You just have to find the best ratio.

FINISH

PERSONALIZE

POST-TREATMENT OF 3D
PRINTED OBJECTS

PERSONALIZE

3D printing has been a geeks and nerds world for a long time. It's only recently that it's become available to artists, stylists and individuals. In order to pay homage to the science fiction side of digital fabrication, the young designer Thomas Thibault has created a telescopic and adjustable Light Saber. Taking the form of a ribbed cylinder, the handle of the toy weapon can be personalized by sliding and screwing different accessories onto it, such as a button, a screen, rings, or even a belt clip. This technique allows anyone to make a unique toy without having to go through the modeling process. However, the main interest of this extremely simple design is the ability to modify it by designing your own accessories while giving them the desired finish (color, texture...). The most impressive element of the saber is definitely the telescopic "laser" that is printed all at once, already assembled. It goes from 15 cm when closed to 86 cm when expanded. This project is a good example of a fun use of 3D printers, especially targeted to individuals. It's also an approach that provides the personalization of objects as an alternative to the creation of multiple "models" with fixed characteristics.

You can download the files for making the Light Saber at the following address:

www.thingiverse.com/thing:572485

Previous page
Light Saber parts
By: Thomas Thibault
3D printed on: Replicator 2
Source: le FabShop
© Thomas Thibault

Left page
Assembled Light Sabers
By: Thomas Thibault
3D printed on: Replicator 2
Source: le FabShop
© ThomasThibault

Right page
Light Saber FabShop
By: Thomas Thibault
Source: le FabShop
© Tatiana Reinhard

POST-TREATMENT OF 3D PRINTED OBJECTS

POST-PROCESSING METHODS

There are many ways to finish your pieces: here is a non-exhaustive list of examples of "post-print" techniques ... It's up to you to invent and discover others according to your needs and desires.

SANDING

The "layer by layer" appearance of 3d-printed objects can be off-putting for certain projects: sanding is very tedious and can turn out to be impossible in certain inaccessible areas of your pieces. Most times we use it for cleaning support traces left on 3D printed pieces, or for fixing points that are a little broken down. Begin with

large-grain paper (P80) then gradually switch to a finer grain (240-400). Be careful, each plastic reacts differently, and colors such as "black" may whiten when sanded. Use a Dremel as needed for more effectiveness.

SMOOTHING WITH ACETONE

This works better with ABS plastic: the material dissolves on contact with the solvent. There are several ways to proceed; some people begin by sanding their piece and passing a paintbrush dipped in acetone over the surface: the dust from the sanding dissolves, filling cavities and smoothing the grooves in the printed layers. This method can be also very practical for gluing pieces together. You cover the two walls with acetone, then place them against each other at the moment when they begin to dissolve. They will bond that way permanently. A simple but dangerous method consists

of dipping the object in a bath of liquid acetone during 1 to 10 seconds and letting it dry. The most effective way is also the most complex; it consists of heating the solvent to use its vapors: be careful, they are toxic, so this maneuver should be done in a well-ventilated area, with all of the necessary precautions. You need a glass container equipped with a lid. Add a small amount of acetone: a few tablespoons or more depending on the size of your piece. With the help of a hot plate, heat the container to around 60°C. Then insert your piece into the inside, on a base where it can be easily

removed and where it will not be in contact with the liquid (think of a handle/rod which will hold the part in the air and help you to retrieve your piece without having to touch the acetone and the bottom of your jar). Wait from 1 to 10 minutes depending on the desired smoothness: if you leave it too long, you risk ending up with a piece that's completely melted rather than smooth... It is therefore necessary to keep an eye on the progress of the operation. It can be repeated as many time as needed. With the help of the rod attached to the base of your piece, retrieve the piece from the acetone chamber and let dry 1 hour before handling it: you can then remove the base and any burrs with the help of a cutter. The use of safety glasses

and gloves is highly recommended. This trick also works with cold acetone, but more time is needed for the smoothing to take action.

The smooth effect is sometimes impressive and provides a finish that resembles pieces produced by injection molding. However, you will naturally lose something in precision and detail.

PAINTING

Pieces made from PLA and ABS are best painted using spray paint. The best results are achieved on pieces sanded with a fine grain.

After sanding, rinse your piece: your surface is now ready to accept a layer of primer or paint. You can then cover the areas of the object that you want to protect from any additional layer. Depending on your needs, work with aerosol or acrylic paint: the painting techniques are the same as for model building, all you have to do is have patience while improving your painting skills.

HYDROCOATING

Little-known but effective, hydrocoating is a process of transferring patterns onto 3D objects mostly used in industry (helmets, car interiors…). The technique can be used on all plastics. Here's a video tutorial of how it's done: **http://www.instructables.com/id/ Custom-hydrographic-design-for-3D-printing/**

You will need hydrographic film (choose your pattern, there are dozens of choices on the Internet), as well as a kit containing cans of primer and

activator. Cover the areas of your object that you want to leave intact with painter's tape. Then, in a ventilated room or outside, making sure to wear a mask, vaporize the primer, keeping a distance of 30 cm between the can and the object to avoid splattering. Delicately place your film, cut beforehand to the dimensions of the bowl, on the surface of the water: wait 40 to 60 seconds for the film to expand. Then vaporize the activator (always wearing your mask) and carefully immerse the object while wearing gloves. Don't hesitate to create handles for your object with tape so your fingers don't hide the object being coated. When the object is in the water, use a stick to remove undesirable film residues from the surface of the water before carefully taking out the object. Once it is dry, remove the tape and varnish to your taste.

There are hundreds of other ways to work on your pieces after printing: drills, laser etching, friction welding, heat deformation… Some people even inflate PLA parts after dipping them in hot water. Feel free to invent your own method!

Photo credit for "acetone" example by Éric Simon - ES Concept.

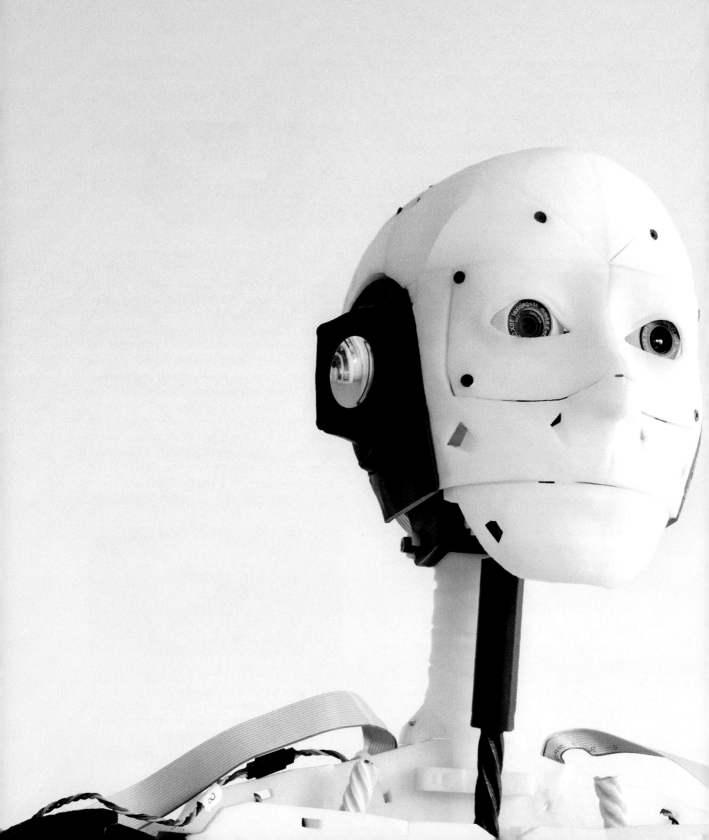

ORDER

THE SECOND LIFE OF YOUR OBJECTS

3D PRINTING SERVICES

PLACE AN ORDER WITH I.MATERIALISE

MODEL OPTIMIZATION
FOR POWDER 3D PRINTING

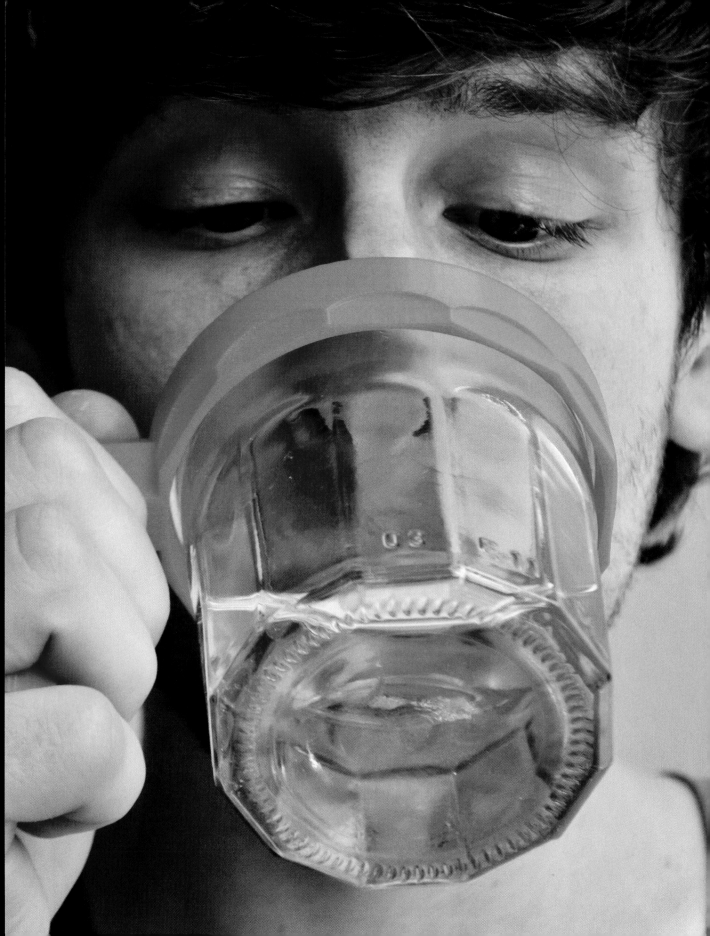

THE SECOND LIFE OF YOUR OBJECTS

Although 3D printing usually uses plastic, the environmental impact linked to this new local production technology can have positive results. The RE_ project (below) consists of making small accessories that are screw on and attach to existing objects in order to give them a new function. Thanks to this idea, an old jam jar can become a bird feeder, a piggy bank, a cup, or even a lemon squeezer.

For the moment, one of the most interesting aspects of 3D printing, from an environmental point of view, is the ability to create replacement parts for repairing defective objects. By repairing objects rather than replacing them, we increase their lifespan. Even if this seems like an obvious fact, our current consumption habits tend towards replacing defective objects rather than repairing them. The replacement or "upgrade" parts for a product can be manufactured locally thanks to digital fabrication, and they remove the costs and environmental impact linked to transportation, packaging, and stocking.

The InMoov project, created by Gael Langevin, is a good example of this close future. His open source humanoid robot is almost entirely made with 3D printing. If a piece breaks, it just has to be printed again. Tomorrow we might be able to order replacement vacuum parts from the local printer after receiving them by e-mail from the manufacturer. Today it is already possible to order your creations and those of others on 3D printing service sites online. Tomorrow... who knows, these services could be on your street! From your personal computer, you can personalize and order a "Bonne Maman" jar handle via the Sculpteo service: **http://bit.ly/1p6vq0y**

3D PRINTING SERVICES

The three main worldwide platforms of online 3D printing were all born in Europe. They offer to load your models for printing, but also to order from creators' stores or catalogs of items, some of which are customizable. These services will allow you to choose among several materials, fabrication methods, and delivery times. These platforms will immediately provide you with an estimate, but the more rushed you are, the higher the price will be.

I.MATERIALISE

This Belgian supplier offers an impressive catalog of 17 materials: you can print in gold, bronze, copper, or rubber, for example. Material specification sheets are useful for helping you to prepare your files knowing the physical constraints of the future object. Moreover, i.Materialise offers the Creation Corner; because a design or a photo does not magically transform into a 3D model, you can hire a designer or customize existing models. If you are using Autodesk's 123D applications, know that you can even retrieve your models by connecting to your Autodesk account. i.Materialise is a subsidiary of the Belgian group Matérialise, the largest 3D printing factory in Europe, publisher of the famous 3D printing management software 3D Magics.

http://i.materialise.com

SCULPTEO

This French 3D printing service is easy to use and very comprehensive. It offers 3D printing in monochrome and colors and in materials such as plastic, alumide, resin, powder sintering, ceramics, and silver. A relevant and practical service for analyzing, retouching, and customizing your files is also available.

http://www.sculpteo.com

SHAPEWAYS

The Shapeways site offers a large collection of different 3D printed objects in a large number of materials and finishes. The platform also lets you offer your own designs for sale.

http://www.shapeways.com

3DHUBS

This 3D printing service concept is highly innovative. Instead of offering factory-like 3D printing, like Shapeways, Sculpteo, and I.Materialise, 3D HUBS developed a worldwide network of 3D printer owners. A customer can go on their website, upload her design, and automatically receive 3D print service offers from the nearby hubs. This system allows faster digital manufacturing at a fraction of the price and, most of all, it's local. You can become part of this community for free.

http://www.3dhubs.com

PLACE AN ORDER WITH I.MATERIALISE

Before starting, go to the materials list in i.Materialise: the maximum size of the printable pieces, the specifications, and even the photos in the overview will help you choose your options and prepare your file. Using the example of a scan created in the "Scan" exercise page 118, we will take a tour of the of the multicolored materials list.

All that's left is to create an account and click on the "Upload your 3D model" button that appears on the home page to get to the file loading page. We will export our color scan in OBJ to get a textured object. Slide your file into the top left area of the page or use the "Upload" button, once again visible after you have chosen your unit of measurement (inches or mm).

After some calculations, your object will appear. A warning might appear: here they are telling us that the model is too big to be printed in the chosen material.

Scan Patrice podium

With the help of the maximum dimensions that the site then gives you, adjust the size of your model using the slider at the bottom left: you can also enter a percentage to make the model larger or smaller, or enter the dimensions in inches or millimeters to get a precise measurement that will be proportionally adjusted to X, Y, or Z. Finally, the "Scale to fit" button will automatically size your object to the maximum possible dimensions.

At any time you can change the choice of material and change your settings.

The price of your order is updated in real time.

If you are satisfied, click "Add to cart" to take you to the ordering and payment pages.

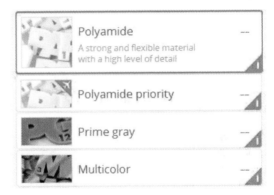

Note that with other materials you will be offered a choice of finishes: once again, everything is indicated in the materials forms (link at the beginning of the exercise).

MODEL OPTIMIZATION FOR POWDER 3D PRINTING

The denser and heavier an object is, the more materials it uses, and therefore the higher the price. In SLS, the fabrication is done in a powder bath: if you decide to make an object that is hollow but closed, the non-hardened powder will still be stuck inside your piece (and unless you want to make maracas, you won't be happy about that). You'll have to make a piece that is hollow (a "shell"), but with a hole in order to remove the powder remaining on the inside. Let's open Meshmixer.

Import your model. Let's first give it a thickness and make it hollow. For a better view of what is happening, change the "Shader" for a transparent texture.

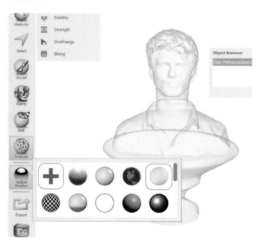

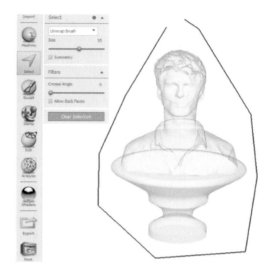

From the "Select" menu, surround the object with the "lasso" tool or use the select all shortcut (Control-A or Command-A).

It will turn orange and new options will appear on the menu. Choose "Select/Edit/Offset" (or Command/Control-D). Enter a negative value in millimeters to create an inner surface.

Click on "Accept".

"Union" will merge your two objects, "Intersection" will only save the area where the two pieces are superimposed, "Difference", which we will use, removes the contact zone between the object, along with the cylinder. Click on "Accept" to validate the boolean and export your hollowed model. With this trick, you can easily save up to 50% in powder 3D printing cost.

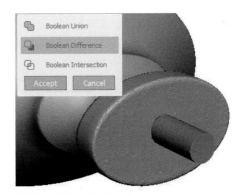

Import a cylinder via the "Meshmix" menu, scale it to a proper size, and put it in the place of your future hole (preferably in a place that will not be visible, like under the base of the bust). Stay in the transparent shader to make sure that you are not touching other walls and inadvertently giving them holes as well.

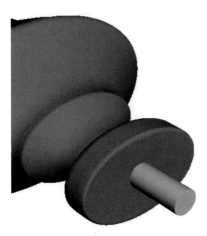

Select your object and then the cylinder by keeping the "Shift" key pressed. You can also enable the object browser in the view tool bar. A new "Edit" menu appears to offer the three boolean options.

CONCLUSION

3D printing, like modeling or three dimensional scanning, is a creation tool among other things, but what most distinguishes it is its potential to create an almost unlimited number of shapes. As with every tool, you need to master it, to take advantage of its abilities, but also know its limits. Prototyping is not the ultimate solution for every manufacturing need. Techniques that are hundreds, or even thousands, of years old such as casting, machining, and blowing, will still stay in our industrial landscape for a long time. The real revolution that these digital manufacturing and design tools provide is that of overall control of ideas, from pencil to prototype, all without leaving your office chair. The private individual, like the creative professional, is enriched and now able, with manual skills and fundamental techniques, to download, create, repair, digitize, and color objects in their daily life.

The other advantage, not to be forgotten, is this new ability to share and collaborate on common physical projects on a worldwide scale: digital objects have the invaluable asset of being easily transferable in order to be modified, improved, and created. There is no need for packaging, stocking, or transportation with objects made locally with a personal 3D printer.

These little exercises have given you a solid foundation for developing expertise in the vast world of 3D, but the key is to practice. By multiplying projects and publishing them on different platforms, you can expand your experience as a maker and designer. Also, manufacturing spaces such as FabLabs, Makerspaces, Hackerspaces, or FabClubs will be an important aid for discovering new machines, new materials, and new processes for enlarging your areas of expertise and feeding your creativity. Personalized manufacturing through online services is only in its infancy, but you can be a part of the revolution that is already going on by becoming an active and prolific creator.

Bertier   Tatiana   Samuel

ABOUT THE AUTHORS

SAMUEL N. BERNIER

Samuel Nelson Bernier is a Canadian industrial designer and le FabShop's creative director since January 2013. He is a graduate of the University of Montreal where he excelled with his exceptional educational journey: Winner of the Lieutenant Governor's Medal, 2010 personality of the year and winner of the Be Open Award in London. In 2012, he was invited by the Autodesk group to be an artist in residence in San Francisco. He wrote the memoir Project RE_, DIY in Digital Age, which followed his journey through the world of Makers and FabLabs. It was this same year that he met Bertier Luyt, a French entrepreneur who invited him to join his business project, le FabShop.

Samuel N. Bernier also taught design at the Milan Domus Academy and gave multiple conferences about the importance of DIY for companies such as IKEA and l'Oreal." In August 2013, the British magazine ICON chose the young designer for its "Future 50" list of creators "shaping the future".

BERTIER LUYT

Bertier Luyt is a professionnal maker-entrepreneur, self-educated, father of two kids. After a career in music, in the early 2000's he turned his career toward designing professional spaces, an activity for which he started using SketchUp. Bertier Luyt's expertise was recognized in 2010 when he gave a conference at Google Sketchup 3D Basecamp about "3D modeling for digital manufacturing", at a time when the first 3D printers aimed at the general public made their appearance in the United States.

After attending his first Maker Faire in New York in 2011, he launched le FabShop, a 3D modeling and digital fabrication studio. Early works include 3D modeling the Palace of Versailles for Google Cultural Institute **www.versailles3D.com**. Bertier has written other books on 3D printing and 3D modeling; he's also a speaker at different conferences such as MakerCon, TEDx, Hello Tomorrow, and Trimble Dimension. His favorite topics are digital manufacturing, self-empowerment, and entrepreneurship. In 2013, he organized France's first Mini Maker Faire in Saint-Malo and created SWF: the first Eco-friendly seaweed 3D printing filament. He is the producer for Maker Faire in France.

TATIANA REINHARD

An illustrator and computer graphics designer first, following a successful career in multimedia visual communication, Tatiana Reinhard, moved by her passion for the image, began a masters degree in Arts and Technology of the Image and became a 3D generalist for video and real time. A compulsive dabbler, passionate geek, and curious about new technologies, she works on the ongoing industrial revolution through the emergence of FabLabs and 3D printing. After teaching CAM at the University of Paris-8, she applied to le FabShop, wanting to test her knowledge as a virtual technician in a digital manufacturing studio.

Today, Tatiana is a designer and creative at le FabShop. She also provides professional training to businesses going into 3D printing.

INDEX